Talking About Mathematics

Talking About Mathematics

TOM BRISSENDEN

with the
Lakatos Primary Mathematics Group

Basil Blackwell

First published 1988
Reprinted 1989
Published by Basil Blackwell Ltd
108 Cowley Road
Oxford OX4 1JF
England

British Library Cataloguing in Publication Data

Brissenden, T.H.F.
Talking about mathematics.
1. Mathematics—Study and teaching
I. Brissenden, T.H.F. II. Lakatos Primary Mathematics Group
510′.24372 QA11

ISBN 0-631-15887-1

Typeset in Century textbook 11/13 pt.
by Colset Pte Ltd, Singapore.
Printed in Great Britain

The Lakatos Primary Mathematics Group

George Ball
Tom Brissenden
Elizabeth Davies
Denise Thomas
Sandra Townsend
Margaret Wagner

Thanks are due to Gaynor Hoare and Pam Parry for their invaluable help at many stages, to the other teachers whose work and ideas are incorporated into this book, and to Brenda Brissenden for constant support through many meetings and much hard work.

Contents

Foreword

The role of language in the learning of mathematics has come to be seen as one of increasing, indeed vital, importance. There are now many research studies in this area, and at the recent international conferences 'Language and Mathematics' has featured for the first time as an independent theme. Teachers in Britain have experienced this surge of interest, first through the Cockcroft Report, *Mathematics Counts* (1982) and more recently through the aims and objectives set out in the Department of Education and Science report *Mathematics from 5 to 16* (1985).

The part played by spoken language, rather than written forms, together with ways of improving classroom interaction, have always been of particular interest to me. During the period 1982–4 I developed a small-scale project based on recording discussions with primary school children about mathematics. The work used ideas developed in an earlier book, whose main focus was the secondary stage. By 1985 it was clear to me that such studies, however interesting, were unlikely to make any practical impact on primary teachers unless they could be widened in several important respects. Of these, the most significant are the need to take account of the mathematical qualifications and attitudes of teachers and the need to make helpful suggestions about classroom organisation – in particular, that it needs to be based on the use of small groups, rather than individuals, as the unit for teaching and learning in mathematics. There is, I believe, a valuable body of expertise within the profession, largely unarticulated, which can help to solve most of the problems if only it can be tapped and shared more widely.

In my own case, help was at hand in the form of an experienced and enthusiastic group of teachers who had followed the Diploma in Mathematical Education course at the Polytechnic of Wales. Their work and ideas during 1985–6 are incorporated into this book and will, I believe, add the essential 'practical touch' that research studies alone are bound to lack. We offer our suggestions as a contribution to the in-service training effort in primary mathematics resulting from the Cockcroft Report.

We were unable to devise any plausible acronym for our group. But our thinking about the nature of mathematics led us to the philosophy of Imre Lakatos, whose famous *Proofs and Refutations* is itself presented as a long debate about the Euler theorem. Lakatos' attempt to set up a philosophy of the process of creating mathematics seemed very much in the spirit of what we were about – hence our choice of name. In the autumn of 1986, the group was absorbed as the nucleus of a teachers' team working under the aegis of PrIME (*Primary Initiatives in Mathematics Education*) on the theme of 'Discussion-based teaching in primary mathematics'. Here we are continuing our work on language and discussion in primary mathematics, in the production of a school-based in-service training pack.

Tom Brissenden
University College of Swansea, Wales

PART ONE:
THE NATURE OF THE TASK

1 The Importance of talk and discussion in mathematics

Discussion and the Cockcroft Report

The Cockcroft Report, *Mathematics Counts*, was intended to be the major influence on the teaching of mathematics in Britain during the latter part of the twentieth century. It is soon clear to the reader that the main themes of the Report, supported by the massive body of research evidence collected in its supporting volumes, are *language and communication, practical work* and *understanding*. Indeed, the Report sees the chief reason for teaching and learning mathematics to be 'because it is a powerful means of communication'. From their earliest days at school and throughout the primary and secondary stages, children are to be encouraged to discuss and explain the mathematics that they are learning. Thus the famous paragraph 243 of the Report *demands* 'discussion between teacher and pupils and between pupils themselves' as an essential feature of mathematics lessons at every level.

The authors of the Report are hesitant about recommending particular styles of teaching. Nevertheless, its recommendations about the kinds of mathematical experience which children should encounter seem to make considerable, and specific, demands on the teacher. These are to do with the teacher's own role in obtaining and conducting talk and discussion, and in planning and organising the mathematical activities within the classroom. Cockcroft itself offers little in the way of detailed guidance on these aspects. Yet the evidence available about past and present practices suggests strongly that great changes will be needed to bring about the situation which the Report demands. The purpose of this book is to help teachers bring about these changes, first, by supplying a helpful framework to guide thinking and training, and second, by providing a number of detailed suggestions to act as models.

The evidence about current practices mentioned above can be found

in a variety of sources. Galton's and Simon's *Oracle* Project carried out studies of both the processes and the products of classroom interactions in nineteen primary schools during 1976–7. Their results led them to conclude that if the stimulating character of teacher-pupil interactions was to be enhanced, then there must be a move away from individualisation as a main approach towards a combination of collaborative group work and class teaching. They regarded total individualisation as impractical – it limited severely the amount of time a teacher could spend with each individual and seriously affected the quality of the interactions that took place. Galton and Simon recommended

> If, on the other hand, the technique of collaborative group work can be extended, this would allow the teacher more time to focus her attention on each of the four to six groups in the classroom, and thus to engage in extended educational interactions more effectively with members of the class as a whole. If this is combined with some class teaching, which also allows this possibility, the general level of stimulation, of enquiry and discovery in the classroom could be raised. (1981, p 180)

More recently, Shuard and Kerslake made extensive tape recordings of teacher-pupil interchanges in mathematics lessons at both primary and secondary level. The overwhelming style was didactic, based on closed questions by the teacher requiring short answers from the pupils. We shall examine this situation in more detail in the next chapter. Their comments echo those made in an earlier research survey by Austin and Howson (1979), where once again it was found that in most mathematics lessons the teacher does most of the talking and few pupils respond. The activities are predominantly of the type 'exposition followed by exercises' and the interaction takes a stereotyped 'question and answer' form. These writers suggest that improvements will depend to a great extent on changes in organisation, with more use made of materials and small groups.

David Lea, a primary headteacher, writing in *The Times Educational Supplement* in 1986, pointed out that few primary teachers had studied mathematics as a main subject (only 1 in 20 according to his own survey). This meant that they were highly dependent on modern, commerically-produced mathematics schemes, where a 'child-centred approach' often meant each child simply working through such a scheme at his or her own pace. This led to class management problems of the familiar sort highlighted by the *Oracle* Project. The busy teacher in such a system has little time to pause for meaningful discussion and so understanding – the fundamental goal – has small chance of developing from the scheme alone. Lea argues that in-service training should start from this situation. Advice is needed, he says, on how to use the interests and personal experiences of children in relation to the mathematical content of familiar schemes. This will surely require the introduction of talk and discussion. Advice is also needed on ways of organising teaching so that the children have sufficient discussion time with the teacher and other children to develop real understanding of the mathematics they are studying.

The comments of members of the Inspectorate, in a survey covering visits to forty Welsh primary schools between 1983 and 1985, strongly support the remarks made by Lea. The Inspectors state that

> In many schools the reliance placed upon published schemes, through which pupils proceed at their own pace, tends to diminish the opportunities for group discussion, and only rarely do small groups of pupils cooperate to attain common mathematical objectives and to improve their problem-solving strategies. (1986, p 18)

Most schools had an adequate supply of appropriate equipment but, because of an emphasis on computation, practical work tended to become an occasional and subsidiary activity separated from the main classwork. It is interesting to compare these criticisms, made in the mathematics section of the Report, with other comments in the section devoted to English. Here we learn that teachers and pupils enjoy easy relationships and few pupils feel inhibited in talking, yet it is unusual for schools to make systematic use of this readiness beyond the infant stage.

> Much discussion is social and fails to impinge directly on the work in hand. Exchanges between teachers and pupils tend to be brief and pupils are rarely encouraged to extend their responses or to pick up in discussion points made by other pupils. The grouping of pupils within classes is seldom used as a means of developing such exchanges. (*op cit* p 8)

In a later paragraph this criticism is extended to the methods used to develop listening skills. Children are encouraged to listen attentively in the early years, and later on commercial materials may be used. But opportunities for children to listen with attention

to other speakers in discussion are restricted, again because of inadequate organisation.

Enough has been said to support my earlier assertion that great changes will be needed to introduce talk into mathematics lessons and so bring about the mathematical discussion recommended by the Cockcroft Report. It is chiefly through talk and discussion that the activities in a commercial scheme become personal experience directly related to particular children, as David Lea requested. Implementing these changes will make new demands on teachers and bring fresh difficulties. To face them, we need to be clear about why the changes are demanded. I would group my own reasons for the introduction of talk into the learning of mathematics under four headings. These are discussed in turn in the next sections.

Talk as a means of improving language skill

Since communication is such a major theme of the Cockcroft Report, it is not surprising to find 'Ability to communicate mathematics' listed as Objective 8 in the Department of Education and Science's *Mathematics 5–16* (1985). This important little book features many other relevant statements, which readers ought to consult in detail. Thus Aim 1.8 discusses 'Working cooperatively', Objective 22 discusses 'Proving and disproving', while Objectives 23 ('Good work habits') and 24 ('Positive attitudes') both contain relevant material to our theme of talk and discussion. Other important sections are Principle 5, on 'Teacher exposition', Principle 6 on 'Discussion' and Principle 12 on 'Organisational arrangements'. All are essential reading for those constructing a modern primary school scheme of work in mathematics. Such a scheme will highlight those aspects of staff development required in order to implement it fully, which should then be the focus of suitable in-service training. This is particularly important in the light of the growing demand for teacher appraisal. Clearly, in view of what was said in the previous section, talk and discussion in mathematics is an aspect which is very likely to need such programmes. The main intention of the ideas and case studies put forward in this book is to provide support for these.

Numerous references in *Mathematics 5–16* show that mathematical talk is felt to be extremely important in its own right. But making it a major element in our mathematics lessons will provide many opportunities to contribute to language development, thus

benefiting other areas of the curriculum. In this way mathematics can make a great contribution to 'communicative competence', a term we shall discuss more fully in Chapter 9. Developments in the recording of work (whether in written or diagrammatic form) are linked closely with those in talk.

It is, of course, part of the teacher's role to inject mathematical terminology and usage into discussions with her or his pupils. This process has to be seen as one of *negotiation of meaning* between teacher and pupils rather than one of *imitation* by the pupils of what the teacher says. In the former, formal definitions evolve as outcomes, not as dogmatic starting points which pupils struggle to accept without being able to appreciate the underlying reasons for the choices which have been made. Precision is clearly an essential feature of mathematics, particularly in calculations and algebraic manipulations, or logical processes. But far too many non-mathematical teachers are ready to accept uncritically the idea that mathematics is some rigidly-defined, unchanging body of knowledge, and that in consequence there is a 'correct mathematical language' on which all mathematicians agree. A better view is that mathematics is a 'way of knowing' – an activity undertaken by people, including learners – where results and procedures are constantly developing. The mathematical philosopher Imre Lakatos attempted to construct an account of this creative process. An indication of his ideas is provided by the following quotation:

> The core of this case study will challenge mathematical formalism . . . Its modest aim is to elaborate the point that informal, quasi-empirical mathematics does not grow through a monotonous increase in a number of indubitably established theorems but through the incessant improvement of guesses by speculation and criticism, by the logic of proofs and refutations. (*Proofs and Refutations* 1976, p 5)

This, then, is 'real mathematics', and its nature can be mirrored to a great extent in the learning process, which it resembles more than many mathematicians would like to admit. If this change comes about, mainly through teachers' use of discussion, then mathematics, by its very nature, could contribute much more than in the past to the general language development of children. We shall try to give specific examples of such developments in Parts Two and Three of our book.

Talk as a means of developing understanding

Theory and research in this area support an intimate link between thought and language and suggest that concept formation and language development go hand in hand. The work of Vygotsky (1971) and Halliday (1973, 1975) is of particular importance in trying to relate language development to the learning of mathematics. There is a growing body of research in this area, some of which will be described rather more fully in chapters 9 and 10. According to these ideas, for the young child language is 'a rich and adaptable instrument for the realisation of his intentions', it is *functional.* The child has, in Halliday's account, to 'learn how to mean' in order to get things done, and so builds up a 'meaning potential'. Thus the acquisition of spoken language is in itself a tremendous boost for further developments, as is the acquisition of reading and writing skills rather later. The schoolchild passes from unformulated to verbalised introspection, and perceives his own psychic processes as meaningful. This, Vygotsky argues (p 91), implies a degree of generalisation and so a shift to a higher type of inner activity, since new ways of seeing things open up new possibilities for handling them. For Vygotsky, 'Word meanings are dynamic rather than static formations. They change as the child develops; they change also with the various ways in which thought functions' (p 124). He concludes that the relation of thought to word is not a thing but a process, a continual movement back and forth; that thought is not merely expressed in words, but comes into existence through them. It seems reasonable to conclude from this that the complicated meanings involved in mathematics will have to be negotiated very carefully with children, as was said earlier; the attempt simply to impose could give rise to much misunderstanding.

Talking – including talking to oneself – will be central to this process of facilitating the development and understanding of mathematical concepts. Typical ways might be through clarification, developing greater awareness, borrowing ideas from other children, extending or refining the available language. The 'conceptual structures' which need to be built up are described in some detail in objectives 10–14 of *Mathematics 5–16*. *Reflection* and *discussion*, I shall argue, following Richard Skemp, are the main components of a teaching and learning process which will attain these objectives. Borrowing Skemp's (1976) terminology, we should aim for 'relational understanding' (knowing why rules work), and 'logical understanding' (being able to explain them to others) rather than the 'instrumental understanding' (using rules without knowing why they work) which results from learning mainly by imitation, as at present. Skemp's work is of particular interest, especially his elaboration of the idea of 'reflective intelligence' and its importance in mathematical thinking. We shall be making use of some of these ideas in the following chapters.

Talk as a means of developing social skills

Mathematics appears to have contributed little or nothing to this aspect of the school curriculum, where other subjects such as English predominate. Yet the potential is obviously considerable – learning to work as a member of a group, articulating complex ideas, acting as spokesperson, giving and receiving criticism, arguing for and against in a logical way and so on. It will often be necessary to have groups of children working cooperatively without the teacher in discussion-based learning. Teachers setting up such a mathematics programme will thus be forced, as were the members of the Lakatos group, to consider the various social skills involved, and how they might be developed. It is rather unlikely that they will just blossom naturally, without some help and planning. Button (1981) has produced materials on the theme of 'developmental group work', but these are mainly directed at the secondary field, and there seems to be little available in the primary stage. Some suggestions for dealing with common difficulties will be put forward at appropriate points in our later examples. A more extended account is given in Chapter 9.

A number of the concerns that surface at in-service meetings are to do with this aspect. First, I would mention *fear of losing control of the class*. This is the worst thing that can happen to a teacher, and relaxing control of the intellectual proceedings seems somehow to be linked with losing control of behaviour. In fact there are a number of 'rules of discourse' to be followed in productive group discussion. Many adults, even those in responsible positions, seem unaware of these and instances of unproductive behaviour are common, for example indifferent leadership, dominance, aggression, withdrawal or task avoidance. A study of group dynamics might be an essential part of both pre- and in-service training in the future! The ability to work as a member of a team is often cited as a desirable quality – in industry, for instance – and might be held to

be indispensable in the teaching profession. Closely linked with the fear just mentioned is the criticism that we often meet from teachers who may say, for example, 'Yes, that was a good discussion with *that* group – but what were the rest of the class doing?' Clearly, it is not enough to demonstrate how to get good work from a small group of children, the organisation of the classwork as a whole must also be covered.

A second problem is *feelings of mathematical inadequacy* on the part of the teacher, much as in the comments by David Lea referred to earlier. A teacher who feels inadequate in this way may be unwilling to allow the talk to stray from the particular foreseen path that he or she feels competent to handle. This situation often develops into a process which is described in the next chapter as 'Guess what I'm thinking' rather than into a discussion. (Note that mathematicians might feel exactly the same if they, like primary teachers, were called upon to teach history or geography!) Of course, this narrow approach may also be adopted by teachers who are well qualified mathematically, but have very rigid views on how the subject should be taught.

Third, there is often *a feeling that discussion is time-wasting*, with nothing to show for it afterwards. Written work, diagrams, models and so on are all evidence that *the teacher has taught*. In fact, of course, the real outcomes of good discussion will influence and appear in such work. But there are several ways of making more permanent records, the most obvious of which is by recording group discussion on cassette. Such recordings can be used in various ways, but some care and training may be needed before making them. More will be said about this aspect in Chapter 4.

Talk as a means of assessment

Listening to what children say during discussion offers the teacher a continuous and detailed means of assessing their understanding and progress. The teacher can respond appropriately and flexibly on the spot. 'Diagnostic discussion' is far more effective in this respect than any other mode of assessment. Again, instances of what we mean will be presented later, in Chapter 3 and Chapter 4. The assessment aspect of talk and discussion can been thought of as unifying the other three aspects that have been described, as in Figure 1.1.

As well as being regarded as a means of assessment, the quality of talk might be assessed in itself. The National Criteria for Mathe-

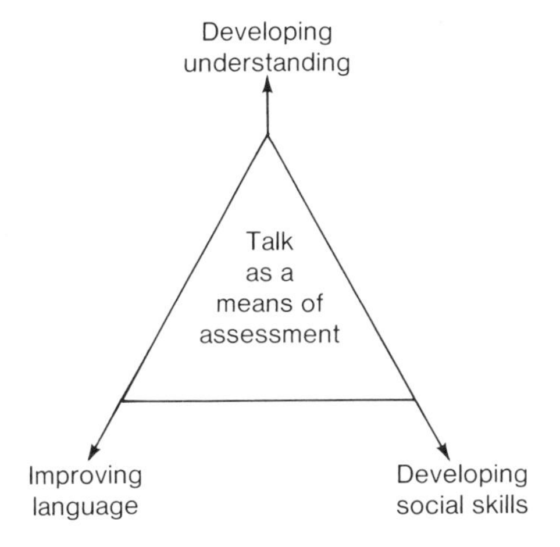

FIGURE 1.1

matics, devised for the new General Certificate of Secondary Education, contain objectives relating to 'mathematical oracy' which may well be directly assessed in the course of time. Much work needs to be done in this area; some initial ideas are suggested in the section on assessment in Chapter 9. The introduction of these formal assessments will obviously make the use of talk and discussion even more vital in primary mathematics.

The what, why and how of discussion

Although I have given four groups of reasons for the use of talk (the 'Why?'), I have not so far attempted to give any clear definition of my use of the word 'discussion'. In fact the attempt to do so can give rise to much debate and indeed some dissension. 'Talk' clearly seems to encompass 'discussion', so let us begin by looking at some kinds of talk that could occur in the classroom.

Teacher talk
- Exposition
- Question asking
- Praising or encouraging
- Controlling pupil talk in various ways
- Discussion with pupils

Pupil talk
- Responding to the teacher's questions
- Responding to other teacher interventions
- Discussion with other pupils
- Discussion with the teacher

The differences of opinion seem to arise over which of these actually count as 'discussion' and whether some can, in fact, occur at all. Thus some teachers deny the possibility of the teacher engaging in true discussion with the children. However, this seems a rather extreme view which I do not accept. In Chapter 2 I shall try to distinguish between conventional 'question and answer' and discussion, and give a fuller definition of the latter. For now, I shall use a definition given by a group of primary teachers at an in-service meeting. Mathematical discussion occurs when:

> Pupils, with or without the teacher, meet together to solve a common problem, or achieve a common goal, by sharing thoughts and modifying their opinions, ideas and understanding.

Only on certain occasions, according to this, will the teacher actually be engaging in discussion with the children – that is, sharing thoughts and modifying her or his opinions and ideas. At other times the 'teacher talk' will take one of the other forms listed above. We saw from the evidence presented earlier that there is a serious imbalance between teacher and pupil talk in mathematics, in favour of the former. The general trend of all our suggestions in this book is towards minimising the 'teacher talk' and encouraging the 'pupil talk'. Exposition, although essential, needs to be kept to a mininum. The teacher will prefer to work through questions which encourage pupils to think and discuss. The form I called 'controlling pupil talk in various ways' is described in more detail on page 40 in Chapter 3; its main aims are to keep the children focused on the problem in hand, to ensure that the interaction is orderly, and to draw out and extend their responses and comments. As before, this form of intervention needs to be kept to a minimum. The teacher will also need to provide frequent opportunities for pupil to pupil talk, in situations which will sustain it without teacher interventions. From the 'What?' and the 'Why?' of discussion we can now turn to the 'How?' – what the remainder of the book is about.

Plan of the book

Our programme is based around a few key ideas, which are developed in a sequence which our own in-service experience suggests is helpful to teachers. So many aspects impinge on the 'discussion theme' that it seems essential to try to introduce ideas in stages, with plentiful illustrations, otherwise readers are likely to feel overwhelmed by the demands which are made. Our ideas will need to be translated into classroom skills, and here again we offer suggestions about how they can be developed in a controlled and supportive way. There are four main stages in this development:

1 The nature of the task
This introductory part describes the background to the task of developing mathematical discussion in primary classrooms. It offers reasons for the greater use of talk in the learning of mathematics and draws on research evidence on present practice to identify the main problems to be overcome by teachers who wish to develop discussion-based approaches.

2 Getting started
Here the main ideas for working with a small group of children are introduced – the kinds of activity (including microcomputer work), the role of the teacher and the roles of the children. Suggestions as to how this might be done without pressure, and how the outcomes might be shared among a school staff, are offered. Most of the key ideas are introduced here.

3 Thinking more deeply
In this part of the book ways of choosing and planning mathematical activities which involve discussion are described in more detail. The teacher's new role is also developed more fully. Suggestions for organising groupwork in mathematics are made, illustrated by several school examples. Groupwork is the vital corollary to the first stage, of course. Finally, some approaches to whole class teaching are given, for those interested in this possibility.

4 Setting in context
In this part the broader aspects are discussed – the contribution of mathematics to 'communicative competence' and school language policy, and a short account of relationships between mathematics and language.

Summary

The key ideas developed in this book may be summarised as follows:

1 Well-planned materials which generate and support mathematical activity by groups of children – including, of course, talk and discussion – need to be introduced into the scheme of work. Clearly, not all mathematics materials will do this, although they

may be effective for individual learning. Examples will be offered, most of them well-known, since we do not wish to 're-invent the wheel' here! I shall use the jargon phrase 'effective mathematical situation' to describe the cases where our materials seem to work well – very often a matter for trial and experiment, since ideas that work well with particular groups of children may not be as effective with others. Unless our mathematical situations really do *generate and support* the children's activity it is highly unlikely that much in the way of discussion will occur. The teacher may be forced to intervene unnecessarily, and perhaps end up doing most of the talking when this happens. Lastly, good ideas that seem to have worked need to be shared in some useful way with other members of the school staff.

2 The teacher needs to change her or his role in working with the children, and to help them to change their own in response – a process I have called 'rewriting the mathematics script' in Chapter 2. It seems particularly vital for the teacher to develop the skills of listening to and observing the children's activity carefully. This careful attention enables the teacher to judge when and how to intervene. Such interventions will usually take the form of questioning or drawing out – all designed to keep the children doing the mathematics, rather than the teacher. Some form of groupwork organisation needs to be developed so that children spend time working cooperatively in mathematics. The organisation should also allow the teacher to spend extended periods with different groups in turn, discussing their work and teaching them as necessary.

Ideas have been borrowed from a number of important sources to help in developing these key ideas. From the Open University's course on *Developing Mathematical Thinking* (EM 235) came the DO – TALK – RECORD framework, which neatly summarises the features one might expect to see in an 'effective mathematical situation'. That is, we should usually observe some form of practical activity (the 'doing'), hear talking going on (naturally!), and see the children employing various forms of recording of their ideas. The Open University team see talking as a kind of 'bridge' between the doing and the recording and their activities are planned on this basis. Their theory will be described more fully in chapter 6. However, I am concerned less with the theory at this stage than with the mathematical activity we can observe. This I think of as an interplay (in an effective situation) between these three forms. The icon shown in Figure 1.2 will be used freely through the text to symbolise this.

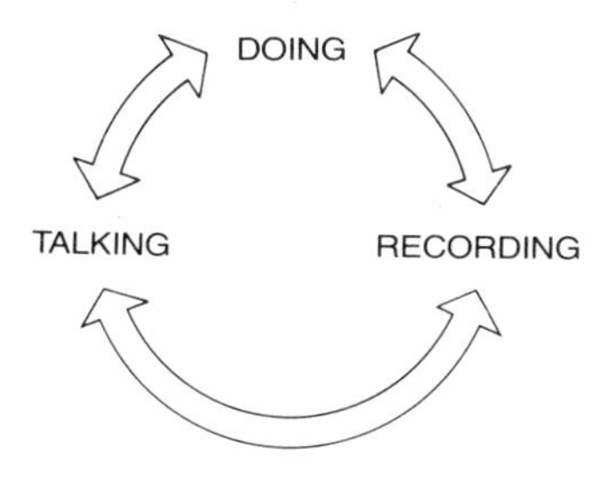

FIGURE 1.2

A further useful framework will be borrowed from Richard Skemp to describe the actual *process of development of ideas* which is going on, to complement the framework of Figure 1.2. But this can wait till chapter 2. For now I shall conclude this chapter by outlining the main features of the teacher's new role, as we envisage it, in Figure 1.3. This diagram incorporates the icon introduced in Figure 1.2, to summarise the children's role. The teacher's role covers four main aspects, all of which have been introduced briefly in this chapter. Each of these aspects will be developed more fully in later chapters.

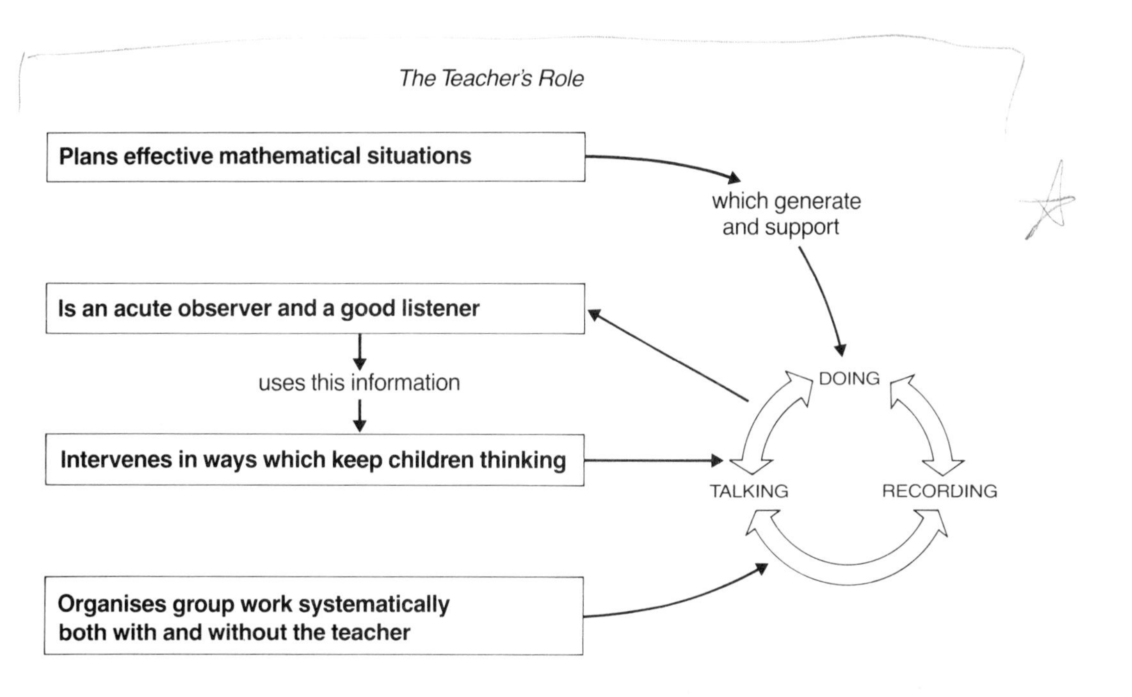

FIGURE 1.3 The teacher's role

2 Rewriting the mathematics script

The title of this chapter is derived from the comment I made in Chapter 1 about the need for the teacher to change her or his role in working with the children, and to help them to change their own in response. I liken the classroom situation to a play with a familiar plot, so that the participants fall easily into their accustomed roles. The 'characters' in the play are, of course, the teacher and the pupils in the class. The 'plot' of mathematics lessons at present, if we accept the various findings described in Chapter 1, contains little or no scope for discussion among the characters, and very little room for pupil talk of *any* kind about mathematics. What we are about can be regarded as nothing less than 'rewriting the script of the mathematical play', for *everyone* concerned.

First, let us imagine ourselves briefly as pupils in a typical mathematics lesson. A possible account might run somewhat on the lines outlined below, but before reading this, why not try writing a few paragraphs yourself, describing such an interlude? I am sure you will not find it difficult to imagine and reconstruct.

A slice of classroom life

> 'What are seven nines?' asks your teacher. He or she wants you to put your hand up, you know, so you expect to be told off if you call out. You shoot your hand up along with twenty others because you think you know the answer. That idiot Jim puts his up, hoping he won't be asked, but your teacher is too smart, and picks him first. Jim has to sweat out a dreadful pause, then the teacher takes time out to comment on his behaviour in unflattering terms. Quite a few hands drop down during this performance, so you wave your arm about hopefully to attract attention and show how keen and clever you are. No luck again – it's bright girl Mary who gets the limelight. 'Well done, Mary!' says the teacher, 'Now – what is six times seven?'

> You decide to sit this one out, but quite a few people call out so you join in anyway. 'Forty-two!' (with perhaps several wrong, but inaudible, answers mixed in!) 'Well, yes,' says the teacher, somewhat reluctantly, 'But I told you before - put your hand up if you know the answer - don't call out.'
>
> You know the answer to the next question, but up shoots Mary's hand so you opt out again. Even if you happen to be picked on, it won't be so bad as with Jim just now - and so the lesson proceeds.

How does your account compare with mine, I wonder? Mine is exaggerated, perhaps, but there is probably a familiar ring to it. We have all spent years in the classroom, learning our parts as pupils in this classroom play very thoroughly. In our own case, later on, we have spent similar periods developing the teacher's role. No wonder everyone knows what is expected of them!. Student teachers seem to be able to role-play their parts as pupils in the classroom play with no trouble at all - even to fitting in minor misbehaviour all-too-realistically at times.

The pupils' roles in this play interlock neatly with the teacher's to produce an *expected pattern*, which is often described as 'question and answer'. Apart from exposition by the teacher, it is evidently the most prominent feature of mathematics lessons, but judging from the *Oracle* findings, by no means peculiar to them. There is very little about this heavily ingrained pattern which resembles discussion, although some people seem to think so, perhaps because it does produce a small amount of 'pupil talk'. If teachers are to break out of this pattern and into one of discussion, they will have to learn very different roles, which will embody *new sets of expectations for both them and their pupils*. We examine what this entails in Part Two, but before doing so, it is useful to study the question and answer pattern in more detail. This will help us to understand more clearly the nature of the changes which are required.

The three-term sequence

The above phrase was used by Flanders, an American researcher, to describe what he found to be one of the most common patterns of classroom interaction. It is a very apt one, as we shall see. But first, I wonder what you think of these comments, which I often meet: 'How can you have discussion in mathematics?' 'Surely there's only one right answer, so either you know it or else you don't?' If, like these people, you do not believe something is even possible, you are not likely to seek ways of achieving it!

I think there is a real foundation for this belief in the one 'right' answer. Indeed, it may be the chief reason for the lack of talk in mathematics lessons, and the very stereotyped pattern of question and answer such as we saw in the previous section. When I ask a question in mathematics, I almost always have a good idea of what makes a reasonable answer – what I shall call the teacher's 'ideal response'. It is hard for me to avoid matching the pupils' replies to my question with this 'ideal response', so I have a constant temptation to comment on everything they say. My comments will usually take the form of an evaluation which makes it clear to my pupils whether I accept their response as 'right' or reject it as 'wrong'. The pupils have learnt that this is part of my role as teacher, and they *expect me to do it*, just as they expect me to ask the questions as another part of my role. So a clear pattern of interaction emerges, as outlined in Figure 2.1.

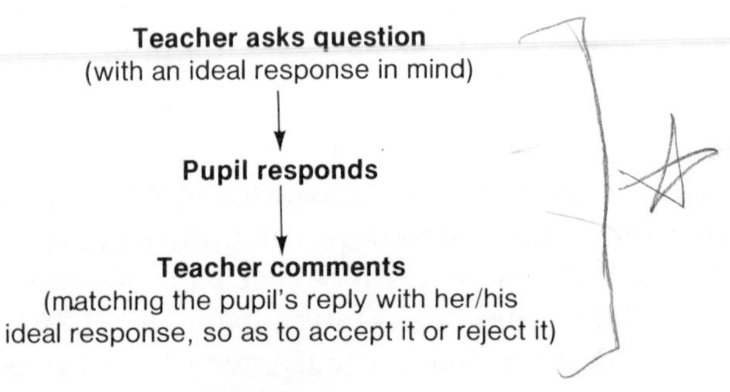

FIGURE 2.1

You can see from Figure 2.1 why the description 'three-term sequence' is an apt one! If the teacher rejects a response, another pupil may be selected or the question may be modified in some way – there are several clearly-describable alternatives which in an earlier book (Brissenden, 1980) I expressed in the form of a flow chart similar to those used in computing. But very often this three-term sequence of utterances is repeated over and over again. The *Oracle* project to which I referred in Chapter 1 used a pattern called an 'exchange', consisting of Initiation – Response – Feedback, or IRF for short. This pattern obviously has similar connotations to the three-term sequence. These interaction patterns resemble a ball-bouncing game in which the teacher is playing against the pupils, as suggested in Figure 2.2.

Notice how this suggests that the pupils speak to the teacher, not to one another. For this reason, although the pupils listen to the

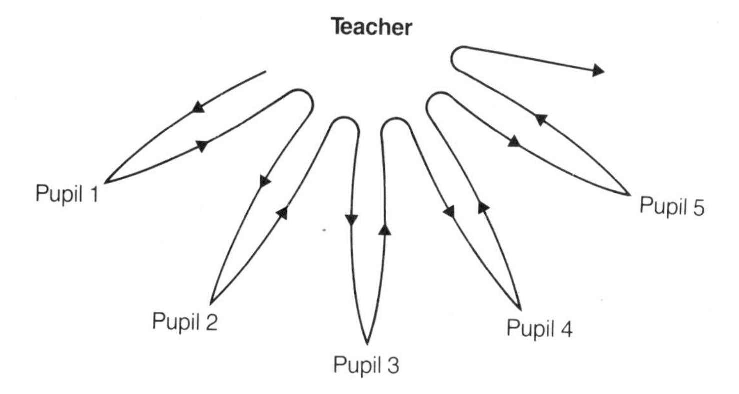

FIGURE 2.2

teacher (usually!), they may not necessarily attend very carefully to the comments of their peers – indeed very often these may not even be audible to all the children, for example when a response is from a child at the front of the class, or one who is quietly-spoken. Research by Nuthall and Church, quoted in Brissenden, *op cit*, and re-quoted in chapter 9 of Quadling *et al*, (1985) showed that children benefited from listening to others' responses, even when not called upon themselves, so this could be an important issue. Nor is the listening of the teacher likely to be of the form necessary for discussion, if it is affected by the process of matching to an ideal response, as in Figure 2.1. No doubt the teacher is gaining information about the extent to which the pupils can repeat back her ideas, and some of the children at least are active. But it is not discussion!

Guess what I'm thinking

The three-term sequences of question and answer flow smoothly enough when the responses are judged to be correct by the teacher. But when they are judged to be incorrect, the interaction can become very messy, and may turn into a process which has been aptly labelled 'Guess what I'm thinking'. Below is a short extract from a lesson to illustrate what I mean (I shall be very sparing with these examples of what not to do!). The teacher is trying with a group of 11 year olds to construct the height of an obtuse-angled triangle, in that awkward case where the base has to be extended and the line falls outside the triangle. Most children experience difficulty in grasping this extension of the idea of 'height', and the present case is no exception. The triangle ABC is drawn on the blackboard as in Figure 2.3(*i*), and the sequence in which it is modified is shown in 2.3(*ii*) and (*iii*).

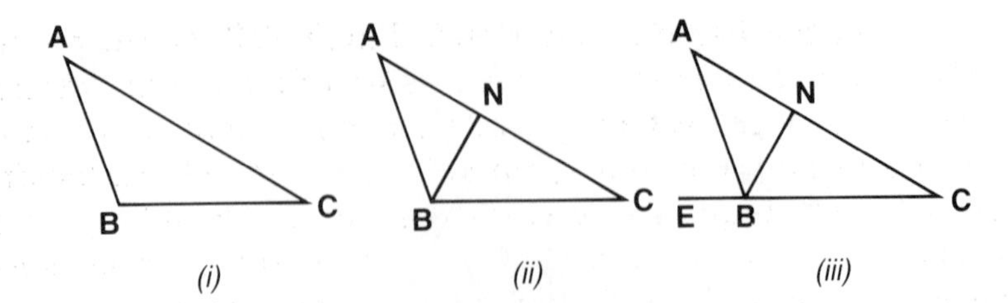

FIGURE 2.3

Teacher: Now, where's the height of the triangle if BC is the base?
Pupil 1: Miss, is it that line?
Teacher: Which line? Name it.
Pupil 1: Line AB, Miss.
Teacher: No, that's not the height, is it? It has to be perpendicular to the base. The base is BC this time. Yes (Pupil 2)?
Pupil 2: I don't think it's got a height, Miss.
Teacher: Yes it has. Where would you draw it from?
Pupil 3: I've got an idea, Miss.
Teacher: All right (Pupil 3), come out and draw it then.
(Pupil 3 draws in the line BN as in Figure 2.3(*ii*).)
Teacher: Er – yes, good, but that's not the line I'm looking for, (Pupil 3). What's the base for your height?
Pupil 3: Er – er – it's AC.
Teacher: Yes, but I want the height when BC is the base.
(There is a pause.)
Teacher: Look – suppose I extend BC like this – can you draw the height now?
(She extends CB to E as in Figure 2.3(*iii*).)

Here a valiant attempt to draw out pupils' ideas and keep them thinking has turned into a guessing game after all. So the teacher is forced into doing the mathematical thinking herself, in spite of her helpful 'prompts'. The difficulty is a familiar one, and one you may have met yourself with older juniors. The mathematical situation here – with nothing more practical than a diagram on the blackboard – badly needs improving if the children are to do the thinking instead of the teacher, in developing their understanding of the meaning of 'height'. Some suggestions to this end will be found later on in Chapter 6.

When clear-cut question and answer turns into the more messy 'Guess what I'm thinking', as in the above extract, the on-the-spot resources of the teacher are usually tested to the full. The responses which they have to evaluate can range from wild guesses, through those which are incorrect or reveal misunderstandings, to some which are correct (in some sense) but not what the teacher is looking for – for example Pupil 3's 'idea' in the extract. Handling these situations successfully is clearly of considerable importance. Bishop has described the various strategies which teachers use, in a list which is reproduced in Kerslake *et al* (1982).

I have observed two effects in such situations – the first rather obvious, and one which took place in the incident described above. The teacher (particularly if he or she is inexperienced) shifts from question and answer to exposition. That is, the children become inactive and the teacher takes over the mathematical thinking. This strategy for coping may be successful in the sense of giving the teacher an immediate course of action to take. It is definitely not successful in the learning sense, however. This is because at a point of difficulty, where in problem solving you would expect the children to be most active and doing the thinking, they do just the opposite! It seems to me that an *effective learning strategy ought to help them to remain active*. This could be done by offering a better 'mathematical situation' to work in, and incorporating opportunities to think and discuss into the lesson plan. In this way we try to avoid the pitfalls of 'Guess what I'm thinking'.

The second effect is insidious and persistent, and I am sure catches us all out at different times. Often we concentrate so much on comparing our pupils' responses with our 'ideal response' (as described in Figure 2.1) that we do not listen to what the children actually say. Indeed, we may even twist their replies around to match our 'ideal', or ignore or reject potentially useful offerings. This happens to the luckless Pupil 3 in the extract. His idea could have been helpful, in suggesting that the triangle could be turned round, but it is pushed aside by the teacher, still searching vainly for her ideal answer. Given this dominance of the 'ideal response' in mathematics, and the tendency it encourages in teachers, as in the three-term sequence, to comment on everything that is said by pupils, it is not surprising that the use of open questions to generate discussion is rare. You can see why we insist that teachers must begin listening to what is actually said, and restrain their urge to cap every statement made by a pupil. Questions remain vital, but are to be used to provoke problem solving, not to search for ideal answers that the teacher already knows.

This is how you do it

In a previous book (Brissenden, *op cit*), in discussing relationships between teaching styles, I suggested that exposition is a kind of 'safety position' to which the teacher can retire when things get too messy in question and answer. In mathematics this position mostly takes the form of explaining 'This is how you do it' to the children – in effect, abandoning the attempt to continue using their ideas. 'This is how you do it' asserts the teacher's authority, both disciplinary and intellectual. Moreover, it puts all the thinking and decision-making safely in one place – with the teacher! But 'This is how you do it' can also arise from a lack of sensitive observation or from the feeling of time pressure – a factor which researchers, or those such as me who are trying to change teaching styles, do not sufficiently take into account. I shall be trying hard to do this in Parts Two and Three! An extract which I have borrowed from Shuard (1986a) illustrates, I think, both these factors at work.

The teacher in this case was working with two second-year junior pupils while the rest of the class got on with other work. Structural apparatus was in use and the children had four tens and two units in front of them.

Teacher: Now from your forty-two it says to take away five; five ones. Can you do that?
Pupil: No.
Teacher: Why not? Why can't you do it?
Pupil: Two ones.
Teacher: You've only got two ones, haven't you? You haven't got enough. Do you remember, when we went through these sums last week, what we said you had to do, if you hadn't got enough?
Pupil: . . . errr . . .
Teacher: What did we say you had to do?
Pupil: You take the . . . errm . . . that . . .
Teacher: You had to borrow. What did you have to borrow?
Pupil: Three.
Teacher: Did you have to borrow three? You had four tens and two ones. What did you have to borrow?
Pupil: Four . . . one . . .

Shuard points out that the teacher urged the pupil to remember the taught algorithm, in which a 'ten is borrowed', and ignored his idea of 'three', which might have been used constructively, by splitting up a ten, for example. But only the preconceived rule is acceptable to the teacher, who does not really listen to what the pupil said and take it up. So the idea of 'three' is lost, and the pupil appears to start guessing, or struggling to recall, the algorithm which the teacher has in mind.

I should think that at this point we will all begin to recall guiltily numerous occasions in which we have acted in a similar manner to the teacher in this extract. 'Teachers don't listen!' is a message that comes to the attentive ear, at every level of the educational system. Teachers on our in-service courses have frequently commented to me, à propos their child studies, 'This is the first time I've really been able to listen to children for years!'. One of the factors at work must surely be the time pressure in teaching a class of thirty children. The *Oracle* project highlighted a teaching style it called 'Individual monitoring'. This style was criticised in terms of the relatively low intellectual level of the interactions that took place, consisting mainly of instructions and information-giving.

However optimistically teachers start out on their careers, they may gradually be *trained out* of dealing sensitively with their children. Sensing other problems developing, and other children needing attention, they learn to press on regardless rather than entering into discussion. It seems clear that any proposals for reform must take account of this, and try to offer ways of reducing these pressures. Hence the factor I have shown in Figure 1.3 as 'Organises group work systematically' is discussed in Part Three, while Chapter 3 contains a suggestion for lessening the pressure in 'getting started'.

At this point I shall list those features which I think characterise discussion, which will be developed further in later chapters. They may be compared with the very brief definition which I gave in Chapter 1. You may well wish to compare them with your own views, and perhaps discuss (!) them with colleagues.

Characteristics of mathematical discussion

1 People speak and listen to one another on an equal footing. Thus there cannot be someone present who comments immediately on everything that is said, as in the three-term sequence just described. Of course, there can be a 'chairperson' or 'group leader', but such a person needs to act so as to encourage and ensure orderly

discussion, not to inhibit it. This may be a helpful way of beginning to think about the new role demanded of the teacher. However, it should only be regarded as a useful analogy, and not pressed too far. There are important differences between it and the teacher's role – for example a chairperson is not expected to teach, or to be particularly supportive of participants in a discussion, he or she is only required to be fair.

2 Participants listen to what is actually said – *not* with some preconceived set of ideal responses in mind. Once again, the difference from conventional question and answer is clear. It is also obvious why I said, in my discussion of key ideas, that teachers will need to become 'acute observers and good listeners', (see Figure 1.3 page 15). Mathematical discussion is likely to demand unusual concentration and so this point is extremely important.

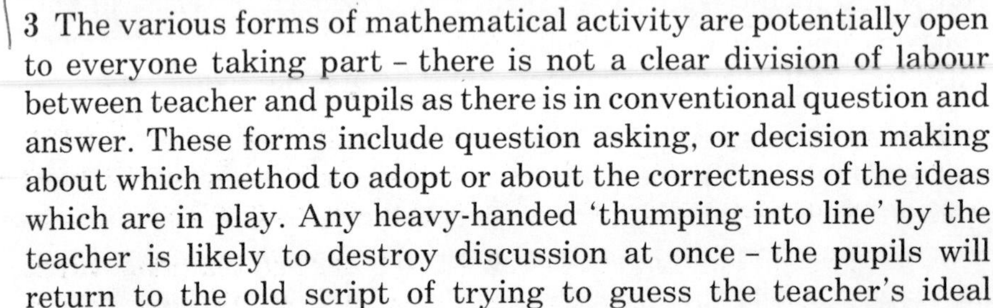

3 The various forms of mathematical activity are potentially open to everyone taking part – there is not a clear division of labour between teacher and pupils as there is in conventional question and answer. These forms include question asking, or decision making about which method to adopt or about the correctness of the ideas which are in play. Any heavy-handed 'thumping into line' by the teacher is likely to destroy discussion at once – the pupils will return to the old script of trying to guess the teacher's ideal response.

Establishing pupil activities with all these characteristics may seem quite a task! It is clear that the teacher is playing a much more restrained role than in conventional teaching, and that the children are doing much more of the thinking and decision making. For this to happen, and for discussion to take place, it is absolutely vital that we try to set up those effective mathematical situations that I introduced in Chapter 1. I shall make this the fourth characteristic of mathematical discussion.

4 Mathematical discussion between a group of children is based on an effective mathematical situation. It will almost always involve the features shown in Figure 1.2 of DOING, TALKING and RECORDING. This means that practical materials will form an integral part of the activity, rather than a separate feature, as in the comments by the Inspectorate reported in Chapter 1. This is what the teacher can observe and hear taking place. The actual process has been described most succinctly by Richard Skemp as one of BUILDING AND TESTING OF IDEAS by the children. Note these two vital aspects – building *and* testing. The situation must enable the children to do both – otherwise the teacher may be compelled to

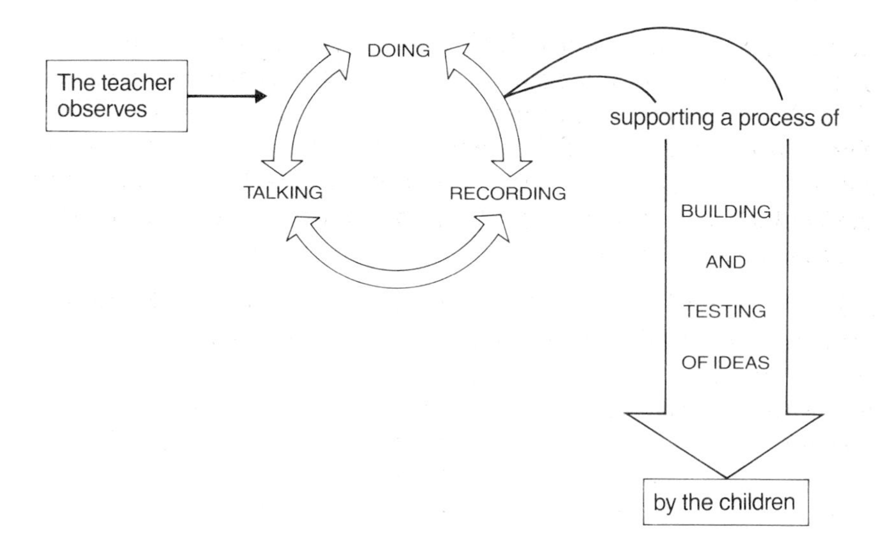

FIGURE 2.4

step in, explaining and telling as in the conventional script. We can now add another section to the diagram of Figure 1.2, showing the process of development which is taking place (Figure 2.4).

This property of enabling the children to build and test ideas must appertain whether or not the teacher is present. Besides this, a number of social skills are required if the children are to cooperate in a mathematical task. I shall develop these ideas more fully in Chapter 9, but briefly children need to

a learn to talk with one another about mathematics – to be willing and able to articulate their own ideas and to comment on those of others;
b be willing to listen attentively to one another;
c learn to take turns – to be willing to give way, 'gatekeep' for others, avoid aggressive or dominant behaviour;
d learn to be supportive – able to criticise or praise without belittling.

As I reported in Chapter 1, the Inspectorate commented that listening skills were developed in primary schools and that children were ready to talk freely. Here, surely, are the opportunities to apply those skills and that readiness in the most constructive of ways, to improve the children's own understanding of mathematics.

The process of building and testing of mathematical ideas of Figure 2.4 can be described in rather more detail in the following way:

Children build ideas
- by thinking about a question (whether posed by them or the teacher);
- by manipulating materials;
- by contributory discussion – *sharing*.

Children test ideas
- by predicting;
- by using materials;
- by comparative discussion – *comparing*.

Exposition and question-asking

I prefer to regard these as two distinct forms of 'teacher talk', although question-asking is at times regarded as a type of exposition, for example by the authors of *Mathematics 5–16*. Under principle 5 (p 38) this document states:

> The purpose of exposition by the teacher is not that everything should be so well explained to the pupils that they become simply passive recipients of mathematics as a body of knowledge. Instead the purpose should be to stimulate and activate pupils so that, as far as possible, they reach the various objectives under their own initiative. Successful exposition may take many different forms but the following are some of the qualities which should be present: it challenges and provokes the pupils to think; it is reactive to pupils' needs and so it exploits questioning techniques and discussion; it is used at different points in the process of learning and so, for example, it may take the form of pulling together a variety of activities in which the pupils have been engaged; and it uses a variety of stimuli.

Exposition remains an essential part of teaching mathematics, but it needs to be integrated into the activity flexibly, and usually in 'small doses' rather than lengthy disquisitions. For example, providing relevant mathematical language at appropriate points, or offering short clarifications of a method or idea, will be very helpful when linked with the children's own ideas arising from discussions. You will find various examples of this approach in later chapters.

Question-asking is a vital tool for the teacher both in provoking mathematical activity by the children, and in the resulting

discussion. It is important to distinguish this from conventional question-and-answer, with which it may superficially be confused. All questions can be thought of as making some sort of demand on the listeners, which forces them to search within themselves. Even very trivial questions such as 'How are you today?' can have this effect! But sometimes teachers make statements which look like questions where it is clear that no response from the pupils is expected. I shall discount these 'rhetorical questions' which the speaker answers himself or herself. Another kind I call 'invitation questions', such as 'Does everyone see that?'. Here, I want to focus on 'demand questions', and particularly in those situations where the teacher is asking them directly, rather than those where questions are posed through a workcard or textbook.

Some teachers make their pupils work hard by using lots of demand-type questions, and very little exposition. This form of teaching is probably what we might term a 'good, traditional maths lesson'. Other teachers tend to use a lot more exposition, and ask fewer, more straightforward questions. This means that they do more of the mathematical thinking themselves than is the case with the previous form. Not only are their pupils considerably less active, but there is a risk of the teacher losing touch with them unless the explanations and general quality of the exposition are exceptionally good.

However, I have no wish to get involved with matters of good or bad styles of teaching. My own impression is that many teachers in fact vary their approach, depending on the children or topic they are working with, so that it is not really very useful to label people with a particular style. Moreover, the notion of 'style' may bring with it aspects to do with personality (such as the teacher's use of humour, for example) with which I am not at all concerned. I shall therefore use the more cautious term 'mode' to describe these various forms of interaction, in any subsequent discussion.

The vital point that I have tried to establish in describing the three-term sequence and in my two examples, is that neither conventional question-and-answer, nor 'Guess what I'm thinking' can help us to move towards the features that I listed as characteristics of mathematical discussion – indeed they are positively unhelpful in trying to 'rewrite our mathematics script'. In the discussion mode, questions are asked in order to *generate problem-solving activity* from the children, not to search for ideal responses. The third term in the three-term sequence disappears, replaced by the activity of our icon in Figure 1.2 showing an interplay between DOING,

TALKING and RECORDING. Teachers will evidently need to push their 'ideal responses' to the back of their mind, to avoid too preconceived a notion of the directions in which talk must proceed, and to restrain their habit of constantly evaluating the comments which pupils make. They will be acute observers and good listeners,  and use this information to judge when to intervene in children's activities, and how. All this has to take place in some form of group organisation, which ensures that there will be a good deal of pupil-to-pupil mathematical talk taking place without the presence of the teacher. These, then, are the matters with which I shall be concerned in this book. At this point, it would be useful to look back at Figure 1.3, and link it with Figure 2.4, to gain an overall impression of how the various ideas are related.

PART TWO
GETTING STARTED

3 Working with small groups

In Part Two I want to put forward the idea of 'getting started' on mathematical discussion by working with a small group of children first. I shall illustrate the idea with a number of extracts from recordings and notes made by teachers working with groups of children, along with suggestions about possible starting points. Without such practical demonstrations no one could take the case for change seriously. Several forms of criticism could be made of these accounts. On the one hand, there is the 'elitist' cry - that some kind of 'super teacher' is implied in the ideas. This is certainly not the case - the activities in the examples were all carried out by good practising teachers, but they would not lay claim to such a label. What is more, in-service trials suggest that other teachers can quickly begin to emulate our own efforts, and many may well better them - this I shall illustrate later on in Chapter 4. On the other hand, there will be various mistakes and failures of insight to be found in the examples, since it is very unlikely that everything will always go perfectly. Such errors could make interesting talking points for in-service meetings. Lastly, the timid or self-satisfied may point to the undoubted difficulties of implementation in terms of time, resources, space and organisation. The ideas and examples put forward in Part Three may help to instil more confidence in such cases. I shall begin with an example of my own.

A Cube Game

Three ten year olds, Joanne, Kathryn and Martyn, are seated round a table with me. They are playing a 'cube game' using 1 cm square lattice paper and a supply of 1 cm yellow and blue cubes. Each child in turn constructs a shape; the other two check whether it is different from all the shapes previously constructed *and have to say why*, with the builder justifying if necessary. The rules for making the shapes (in this particular game) were: a base of five yellow cubes must be placed inside a 3 $\times$ 2 rectangle drawn on the lattice, then

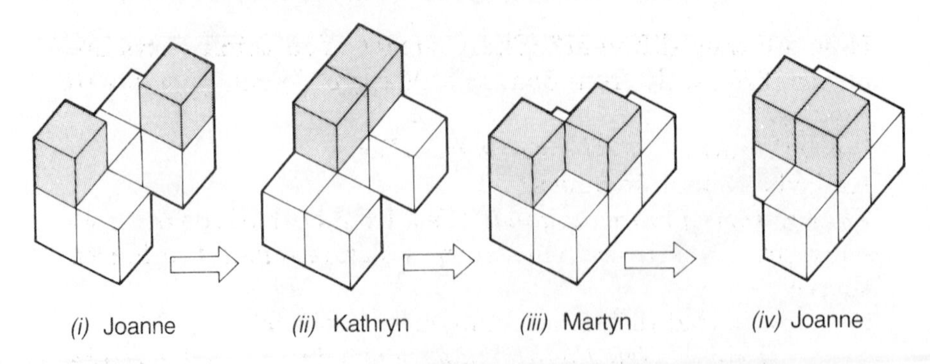

FIGURE 3.1

two blue cubes placed on top of the base, all the cubes having to be 'fitted together properly'. After a short practice session, Joanne made the first shape shown in Figure 3.1 and Kathryn completed the second (ii). The extract below follows the talk through the first round.

Ah, now, Joanne, is Kathryn's shape different from yours?
Ye-es.
In what way?
Cos it's got the same shape in yellow as me but the two blues are in a different place.
Are you happy about that Martyn? Kathryn's is different? . . . Now it's your turn, I think, isn't it?
(Martyn builds his shape.)
Hmm – is it different from yours, Kathryn?
Yes.
How's it different?
There are still five yellow ones but then he's left the space in a different place and, um, my two blue ones are touching rather than leaving a space in the middle.
Joanne – is it different from yours?
Yes.
In what way?
Cos, erm, his yellows are – he did five as well but, erm, I've – I've left the space in a different place and the two blues, erm . . . huh, erm –
There's something different about the blues, isn't there? Now Kathryn's are touching, aren't they, her two blues? . . . Now yours are apart like Martyn's but they're a bit different, aren't they?
Yes . . .

How are they different? (Short silence) You think yours are placed differently from Joanne's, Martyn? Would you like to say then?
Yes, mine are . . . straight to each other.
And what about Joanne's?
They're sort of more diagonal. (*Yes*, from both of the others)
All right . . . is it Joanne's turn? (Joanne builds the fourth shape)
OK, now is that different from yours, Kathryn?
Yes . . . Joanne's left the first square missing . . . and, um, the two blue ones are placed in a different way from mine.
Now what about yours, Martyn?
Mine's different from hers because the shapes – the yellow shapes are the same but the other way round, the blues are different.
We've got a little problem here – are we going to count this base – let's just look at the bases for the moment – are we going to count that as really the same as Martyn's or not?
I don't think it is. (Decisive agreement from Kathryn and Martyn as Joanne says this) *If I turned it round then the square missing would be there and not there.* (Joanne points as she says this)
Ah, that's a good point, isn't it? Shall we turn it round? What do you mean – like this, Joanne, there?

Remember, in pondering the last part of this extract, that the children were seated around a table, with their own shapes in front of them on a single large sheet of squared paper, so that their view was not that of Figure 3.1. Notice how the initial one-word statements of the children are expanded as the game enters its second round, so that my prompting is no longer needed. Joanne fumbles for words in trying to distinguish her shape from Martyn's, but manages to steal the phrase 'left the space in a different place' from the more articulate Kathryn. The trio unites over Martyn's comparison of his and Joanne's blue cubes – 'They're sort of more diagonal' – although Joanne had been unable to articulate either comparison. But immediately afterwards Joanne takes the decisive lead in explaining clearly and succinctly why the base of her second shape is not the same as Martyn's – 'If I turned it round then the square missing would be there and not there.' The game proceeded in this manner through seven rounds, in which twenty-one shapes were constructed, which the teacher praised as a *group* achievement. Obviously it becomes more and more difficult to produce distinct shapes and justify them in later rounds. The children developed a close involvement in the task, and clearly felt a considerable challenge to go on, with great excitement at several stages.

This extract illustrates the point which I made in Chapter 2 about having an 'effective mathematical situation'. The children can both build their ideas and use the shapes to test them, as in Skemp's formulation. They make their own judgements about how they are getting on, thus minimising the mathematics I need to do. You can see how both DOING and TALKING are going on, while the shapes themselves act as a kind of RECORDING of what has taken place. A little later you will see how I introduced a more permanent form of recording in an extension activity. The rules for this particular game (based on trials with another group) contain just enough constraints to confront the children with important decisions, such as the one about mirror images in the extract, but still leave sufficient creative scope. Later on in the game, a similar decision had to be made about 'arch shapes', as we called the first two bases in Figure 3.1. The game format produces the talk, through the rules about each player's new shape being checked by the others. This idea (yet another from the Open University's EM 235) can be used to transform many other situations into ones which satisfy the DOING, TALKING, RECORDING icon of Figure 1.2.

One can foresee that certain kinds of discussion will be 'precipitated' by the careful design of mathematical situations, but not, of course, exactly when or how – as with the mirror images on this occasion. On other occasions language developments happen in an unforeseen way.

To illustrate this last point, let us hear a later round of this game, in which I have to explain my use of the word 'scope'. There are thirteen shapes on the table at this stage, so life has got complicated for the players. Martyn makes an innovation, with the first shape shown in Figure 3.2.

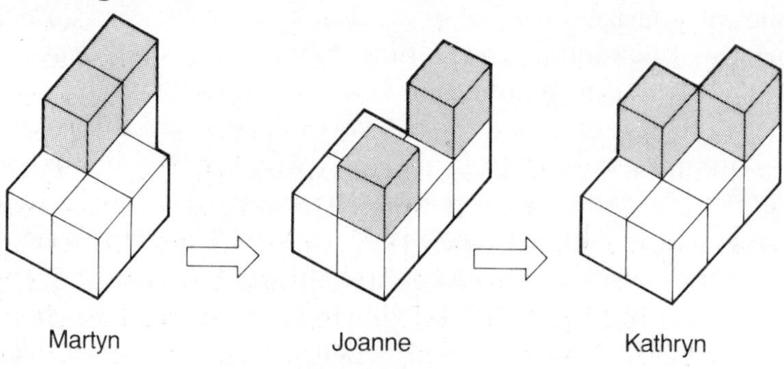

FIGURE 3.2

Ah – yes! Nobody's thought of that yet, have they? You claim that's different, do you Martyn? In what way?
Because my top blue is on the edge – the very edge.
It's on the cube that's sticking out, isn't it? Yes – nobody's thought of that yet, have they? . . . Joanne, come on then . . .
(Joanne makes the next shape, with a little prompt.)
Ah – you could try Martyn's idea, couldn't you – how about putting a blue on the one that's sticking out and then seeing what we can do – ah, yes, that's clever! Is that different from yours, Martyn?
Because her base is the other way round . . . and the blues are apart.
Yes, I think we've got some scope here – right, Kathryn –
What's 'scope'?
Erm – more opportunities – more possibilities – more scope.
Right – I didn't know what it meant. (Kathryn laughs.)
That's what it means – Martyn's idea there has given us quite a lot more scope, I think, because I think we've got a few chances to make some more, haven't we, by putting our blue on the one that's sticking out?
(Kathryn makes the last shape shown in Figure 3.2)
Yes, I like that – Kathryn has borrowed your idea as well, Martyn. Yours is different from Martyn's first one then, Kathryn – in what way?
. . . the yellow – one of the yellow blocks is in a different place by there and my two blues are diagonal, Martyn's aren't.
All right – now, Joanne's got one a bit like that – is yours different from Joanne's?
Yes.
Her yellow base looks the same as yours doesn't it? . . .What about your blues, then?
Joanne's blues aren't touching – mine are – on the corners.

So the children here learnt about the meaning of a new word – with that meaning practically demonstrated by events in the game. As a follow-up to this game, I got this group – who seemed to get on well together – to help one another draw various cube shapes on triangular 'spotty paper' (ie an isometric lattice). Various types of grid and lattice sheets are produced commercially – these sheets can be very helpful both as practical materials and for recording purposes. An interesting booklet about them has been written by Hames (1982). First I got the children to try drawing a single cube, then three cubes in different arrangements, then four cubes making a 'corner' and finally a shape made from six cubes. I eavesdropped

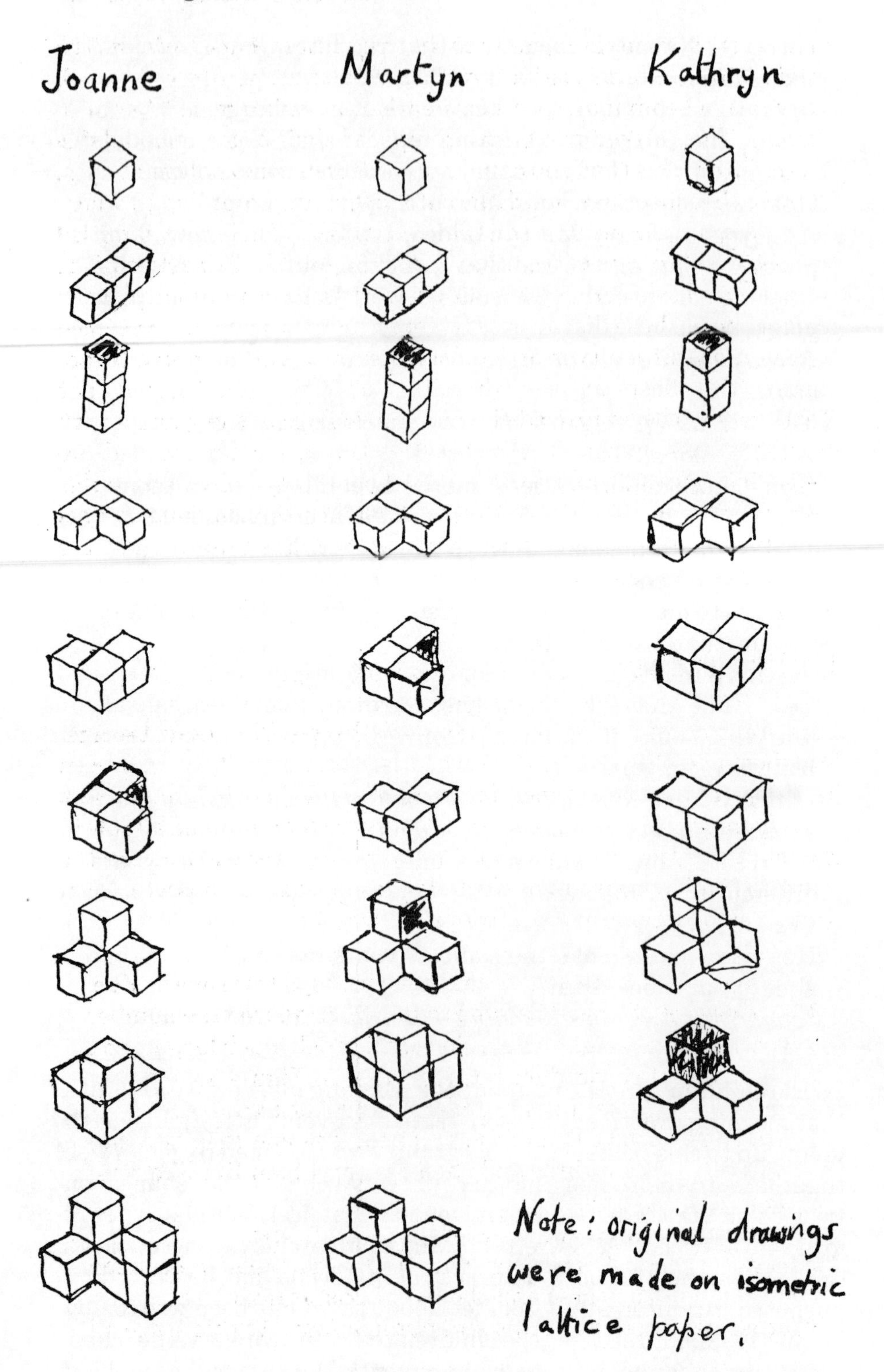
Joanne
Martyn
Kathryn
Note: original drawings were made on isometric lattice paper.

FIGURE 3.3

on this activity, intruding only to set them a new shape to draw. The previous game seemed to have helped them to cooperate well - such cooperative behaviour, as we saw earlier, is an important factor in success. They talked about and compared their drawings, and discovered from this that you could draw different views of the shapes. This was because they had different viewpoints according to where they were sitting around the table, of course. Their discovery led them to try to draw the different viewpoints themselves. The recordings of this activity are shown in Figure 3.3, in which you can trace some of these discoveries. Interestingly, Joanne, who seemed to have most difficulty in articulating her ideas earlier, proved quite clearly the best at the drawing activity. She helped out Kathryn - who had proved the most articulate in the game - with several of the shapes. The difficulties the latter experienced are visible in Figure 3.3, and you can see where Martyn also had trouble with his corner. Joanne was clearly leading the thinking in this activity, which generated a rash of isometric drawings around the class for several days afterwards.

Unifix steps

This group consisted of five children, Catherine, Kimberly, Leanne, Matthew and Steven, aged between four and five. They had been introduced to the numerals 1 to 5, and could make sets corresponding to these numerals. They were seated round a table in the mathematics corner, and *Unifix* cubes were placed in the middle. (*Unifix* cubes can be slotted together to form rods which can broken apart again in any appropriate way.)

Teacher: I want you to make some steps from one to five.
(Matthew set the pace for the group, and produced the number steps in order straight away. The others took much longer, and after a time their productions looked as in Figure 3.4.)
Kimberly: Miss, I don't understand what to do . . .
Matthew: Look Kimberly, you need one and then two and then three and then four and then five like this.
Kimberly: Oh yes.
(She then produced the same steps as Matthew.)
Steven *(to Catherine):* You need to put four in between the three and the five. Look at mine.
(Catherine agreed with this and put in a four. At this stage all

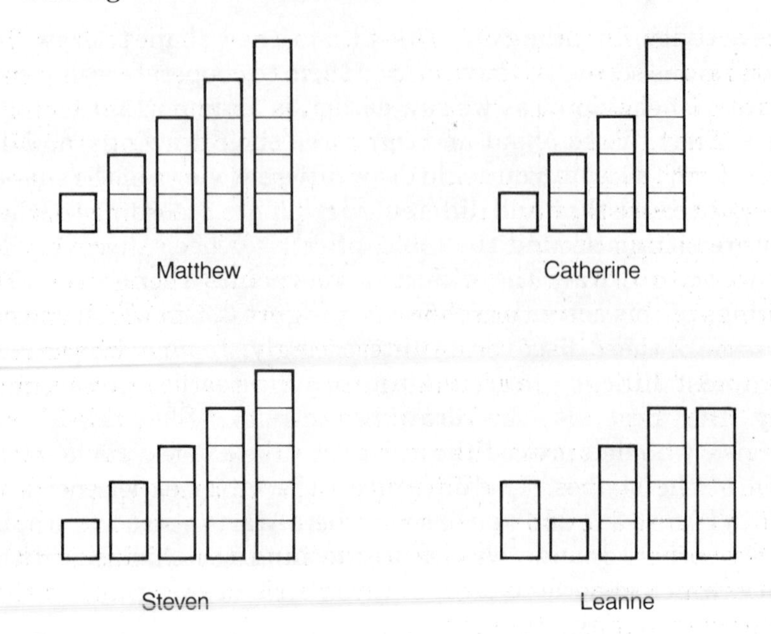

FIGURE 3.4

the children except Leanne had produced the number steps in order from one to five. They had worked together and helped each other with encouraging results. The teacher had not intervened, but now decided to bring in Leanne.)

Teacher: Well, you've all worked very hard to make your steps. Let's have a look at Leanne's steps – what do you think of those?

Catherine: I think they are lovely.

(The others): They look nice.

Kimberly: You need two to come after one and then three to come after two.

(Leanne rearranged her set of rods from Figure 3.4 to Figure 3.5(i).)

Leanne: One, two, three, four, five, six. (She pointed to each rod in Figure 3.5(i) in turn as she said this – quite rightly, since there *are* six rods!)

Kimberly: No, that's not right, you've got one, two, three, four, four, four.

Teacher: How can we change Leanne's steps to match what she has said?

Kimberly: You need one, two, three and put one from the four

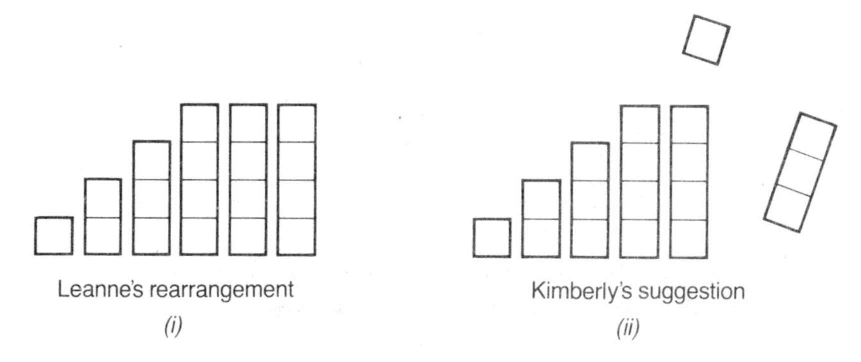

FIGURE 3.5

onto three to make five and then you don't need those three.

(Kimberly's suggestion is shown in Figure 3.5(*ii*).)

Notice the very limited involvement of the teacher in this extract. At no time in the session does she make any comment about the correctness of the children's ideas, although it might have been better to look at each child's production when she intervened, rather than immediately use Leanne's. Once again the practical nature of the mathematical situation helps the children to make their own judgements. Kimberly is uncertain about what is expected of her at the outset, but is helped by Matthew. It is clear that this was not merely copying because later on Kimberly confidently takes the lead in this group in helping Leanne to rearrange her rods.

This extract clearly illustrates the points that I listed at the end of Chapter 2. There I said that teachers would need to push their 'ideal responses to the back of their mind', avoid preconceived notions of the directions the activity must take, and restrain themselves from constantly evaluating the pupils' responses. The extent to which you find you can do this is a sure guide, I think, to the effectiveness of the mathematical situation you are using. Both the cube game and the *Unifix* material worked well, I felt, in this important respect. An interplay between doing, talking and recording is clearly in evidence on each occasion, and the children are able to build and test their own ideas, as Richard Skemp demands.

What of the teacher's role? When she or he is present in the group, observing and listening will always be vital. Asking questions in order to find out what has been happening will often be another kind of 'teacher talk'. The teacher tries to draw out an account from the children of their ideas, to get others to comment on what has been said, or direct their attention to some feature that might help. I have found it useful to draw a distinction between intervening in the mathematical activity itself (for example by asking a question, explaining or prompting) and intervening to help keep the group functioning smoothly and focused on its task. In the latter case, the teacher will draw out children to help or explain, or perhaps criticise – but carefully tries to avoid doing any of the mathematics for them. I shall introduce two more terms to describe these forms, namely the teacher's *mathematical role* and *procedural role*, respectively. My own procedural role is very much in evidence in the cube game, for example, where I keep reminding children of their turn, or asking them what they think about the newly-created shape – 'Are you happy about that Martyn? Kathryn's is different . . . Now it's your turn, I think, isn't it?' In the 'Unifix steps' case, the teacher asks the group 'Let's have a look at Leanne's steps – what do you think of those?

When and when not to intervene became a major talking point in the discussions of the Lakatos group. 'When in doubt, hold back', seemed to these teachers the soundest guide on most occasions. It can be quite surprising how children will come out with ideas and methods, if the teacher waits a little. Such ideas will not be heard unless we can exercise restraint. Certainly mathematical interventions should be kept to a minimum, although procedural interventions will often help to move things along, particularly in the early stages of group work, or in larger groups. As numbers increase from two, three or four up to ten or twelve, so clearer leadership will be needed. Working with a full class is particularly challenging, and ideas for this will be developed in Chapter 8. In order to make these important judgements about when and how to intervene, we shall have to train ourselves to become *attentive listeners to what is actually said*, and *acute observers of children's actions*. The very act of trying to make records of classroom events has helped the members of the Lakatos group to become better listeners and observers, and also more aware of children's use of language, as well as of the potential of the various activities in their mathematics schemes. In my next three accounts I shall try to show more unforeseen – and perhaps surprising – things turning up for the teachers adopting this more flexible approach.

What's the difference?

A group of third year juniors (nine year olds) were working on the 'difference' aspect of subtraction. They were used to the formal algorithm and could apply the decomposition rules quite reliably. A group of five – Kevin, Lee, Sean, Helen and Louise – were working on the same problem; 'Find the difference between 56 and 29'.

Teacher: What do we have to do, Sean?
Sean: Find the difference between fifty-six and twenty-nine.
Teacher: What do we have to do first?
(No reply from the children.)
Teacher: Well, how would you start, Helen?
(Still no reply!)
Teacher: Does anyone have any ideas about where to start?
(Still no response was forthcoming.)
Teacher: Well, if Louise has fifty-six pence and Kevin has twenty-nine pence, who has the most money?
All children: Louise!
Teacher: Right, so do you think it's a good idea to start by finding which number is the biggest?
All children: Yes.
Teacher: Well, let's write the bigger number first, fifty-six . . . now write the smaller number underneath . . . we have to find the difference between fifty-six and twenty-nine – does anybody have any ideas?
Lee: It's one to thirty, and it's twenty to fifty and it's six to fifty-six.
Teacher: Say that again Lee, and write down what you are saying so that we can follow you.
Lee: Start with twenty-nine and go one, then it's thirty. Then it's twenty from thirty to fifty, and then it's six from fifty to fifty-six.
(As he spoke, Lee wrote down the sequence of numbers shown below.)

29	1	30
30	20	50
50	6	56

Teacher: Did you all understand what Lee did?
Children: Yes.

Teacher: Can you all write down Lee's way?
(The children all tried this but their recordings looked quite different. However, they all used the counting-on method that Lee employed – no sign of the decomposition algorithm so laboriously practised in previous lessons!)
Teacher: So what is the difference between fifty-six and twenty-nine?
Children: Twenty-seven!
Teacher: Let's try some more!

In the 'Touching corner'

A group of four-year-old infants, Victoria, Gareth, Dale and Tanusuree are working together in the 'Touching corner'. Language to do with the sense of touch is introduced by a game using two hoops and a box of toy animals. The teacher aims to get the children to sort two sets of animals (furry and not furry) and do some number matching using the sets. The children take turns to draw an animal from the box without looking, describe it from touch and allocate it to a hoop.

(Gareth begins by describing the toy animal he has drawn out.)
Gareth: It's hard and spiky.
Tanusuree: It could be a dragon.
Victoria: Don't show it to us, Gareth, we've got to guess what it is . . . I've seen it now, it's a dinosaur.
Teacher: Which set do you think it belongs in?
Victoria: The blue hoop because that is the one for hard animals.
(The children agree with Victoria's criteria for sorting and put Gareth's animal in the blue hoop. Next it is Victoria's turn.)
Victoria: It's soft with a hard head.
Gareth: I think it's a tiger because it's soft. Victoria, is it a tiger?
Dale: Oh let's see what it is – I don't know –
Tanusuree: I know what it is – it's a monkey – I saw it.
Teacher: Which set do you think it belongs in?
Victoria: The yellow hoop because it is soft.
(She has decided that the two sets are to be 'soft animals' and 'hard animals', whatever the teacher might have had in mind in placing the animals in the box beforehand. Now Dale takes his turn.)
Dale: I'm going to pick a hard one like Gareth's. It's got

lots of legs – one, two, three, four, five, six, seven, eight, nine, ten, eleven, and it's hard.

(Dale counted as far as he could in describing his toy.)

Gareth: I saw a toy – a big worm with lots of legs – I think it was that one I saw.

Teacher: You're not sure of the name are you – show them, Dale.

Gareth: It's a centipede!

Victoria: It's really hard.

(Gareth and Dale were now playing with the animals, being particularly interested in the centipede. The teacher draws the group together.)

Teacher: Which set do you think the animal belongs in?

Dale: In there because it's hard.

(Tanusuree now describes her animal.)

Tanusuree: It's soft and it has stripes.

Dale: Is it a spider? There is a spider in there.

Tanusuree: No, it's not a spider.

Victoria: I know it's a tiger and it goes in that set – I'm right.

Teacher: Which set?

Victoria: The set of soft toys.

Teacher: You've worked very hard to make these sets, can you tell me what you have?

Dale: Yes, a set of hard ones and a set of soft ones.

(The other children agree with this. Gareth begins afresh.)

Gareth: It's hard and it's got one, two, three, four, five, six, seven, eight, nine legs.

Tanusuree: Is it a spider or a busy bee?

Gareth: Yes, and it goes in that set with the other hard ones.

Teacher: Look at your sets very carefully. What can you tell me about them?

Dale: I think there's more hard ones, one, two, three, and there's one, two in that set.

Victoria: Yes, there's three in there and two in there.

(Tanusuree is not so sure about this – she points to the soft set and counts 'One, two, three, four, five' without matching finger to objects.)

Gareth: No, Tana, one, two, three.

The teacher's comments on this activity were as follows:

1 The game situation proved very enjoyable for the children.
2 They worked together as a group, cooperating to produce

the two sets. These were not the attributes I had in mind at the start, but I followed the children's lead.

3 The game helped the children to develop discussion skills – listening to each other, describing things so the others could guess what they were.

4 I was able to assess their use of language and mathematics – for example the need for far more counting activities with finger matching, particularly with Tanusuree.

5 I could plan the continuation activity eg draw sets and match them one to one, or 'How many animals altogether?' with the addition of two sets.

6 As this group were not in my class, the activity allowed me to get to know the children better.

7 I could try the same activity with older children, expecting a more sophisticated use of language. The mathematics should be of a higher level, for example, perhaps the intersection of two sets, since there is a toy that is hard and soft, the monkey, and use of pictorial representation.

A further comment might be to do with Gareth's response to Tanusuree's question 'Is it a spider or a busy bee?' 'Yes', he says 'And it goes in that set with the other hard ones'. This exchange is extremely interesting. First we note that Tanusuree at four can pose a question with alternatives. What sort of responses are appropriate to such questions – and how do we learn to make them? Gareth knows that his animal is one of Tanusuree's alternatives, but evidently does not realise, as an adult might, that a mere affirmative is inadequate. Tanusuree has learnt to ask this type of question, and must have got informative replies from parents or teachers. Unfortunately our transcript can tell us nothing about her state of mind following Gareth's reply. Note that we can reconstruct what the two sets were up to this stage – the 'hard set' contains a dinosaur and a centipede, while the 'soft set' contains a monkey and a tiger. But Gareth's response leaves us, like Tanusuree, still wondering! Questions with alternatives can often be disconcerting to adults. Unfortunately I am prone to this habit, I understand – for example, I will ask my wife 'Do you want a coffee now or do you want to go on working?' 'Yes – No!' splutters Brenda. Now what should I do – make some coffee or wait till later? Perhaps you can think of your own examples. Do teachers pose questions like this during their lessons, I wonder?

What does 'And' mean?

This group of top juniors (ten year olds), Wendy, Stephen and Matthew, taped one of the most enjoyable discussions I have had. From it I have extracted an interlude in which we turn from the game on which the talk was based to discuss the meanings of the different uses of the word 'and'. I used a mathematical situation based on another idea from the Open University, called *Trio tricks*. (You will remember that I warned you that most of the ideas in my examples would be familiar.)

I have three cards with the numbers 12, 4, and 3 written on them, and the object of the game is for the group to construct as many correct statements as it can about them. To get the children to talk, I use the idea of taking turns to write and read out each new statement, with the others checking and verifying its correctness. No apparatus is in use this time – the game demands the ingenious use of mathematical language to express knowledge. At the same time the teacher is able to diagnose and assess understanding. The group has managed to write sixteen different statements when, after a hint from the teacher, Wendy writes down and reads out 'Twelve is greater than three and four'. The teacher intervenes to find out 'What this means'. The talk continues in the following fashion:

Stephen: Twelve is divisible by four and three. (This provoked laughter, arising from the use made of the commutative principle at different points previously.)

Matthew: Twelve is greater than four and three.

Wendy: Four and three are smaller than twelve.

Stephen: Four is smaller than twelve and three. (He first writes 'and', but after some discussion replaces it with a + sign.)

Matthew: Four is smaller than three plus twelve. (He benefits from the previous discussion.)

Wendy: Three is smaller than twelve and four.
(But she writes it with a + !)

Teacher: Let's just leave the game for a second to talk about this 'and'. If I said four and three is twelve – four and three makes twelve, what would you think?

Stephen (amid laughter): We'd think you'd gone berserk!

Teacher: What about four and three makes seven?
(Jointly): Yes – could be true – yes!
Teacher: What about 'and' then? . . . 'Two and three' – complete that sentence.
Matthew: Five.
Wendy: Makes five.
Stephen: Yes, two and three is equal to five . . . um, are larger than one.

(Using two meanings of the word 'and' in the same statement, but I missed it until I listened to the recording later!)

Teacher: But you were using it to mean something different just now. Who wrote this sentence at the top? . . . OK, read out what you wrote, Wendy.
Wendy: Twelve is greater than three and four.
Teacher: Now, what did you mean by that?
Wendy: Those numbers, three and four.
Teacher (relentlessly): Now, what did you really mean by that sentence, Wendy?
Wendy: Well . . . twelve is greater than – both three and four added together.
Teacher: That wasn't what she said the first time, was it, Stephen?
Matthew: No –
Stephen: No, she meant that twelve was greater than three and twelve was greater than four.
Teacher: Was that what you meant the first time, Wendy?
Wendy: Yes – when I'd written it down.
Teacher: And now you mean something else! (Much laughter, including Wendy.)
Teacher: You've said something that means two different things!

(More laughter.)

Teacher (as the laughter dies down): Could it mean two different things?
(Jointly): Yes – yes!
Stephen: It can – if you put a plus sign there –

The group went on to identify these meanings clearly, as well as confirming that both assertions were true, before continuing with the game. Interestingly enough, the use of 'and' cropped up several times more, with Stephen playing a key role in the discussion. The discovery of ambiguity was perhaps as important as anything else that happened in the game, but I had not, of course, foreseen it. How would you have tried to handle the discussion at the point where we branched off from the game? Amongst the things I learnt about the group were its *a* considerable fluency in terminology such as 'is the product of', 'is divisible by', etc.; *b* knowledge of 'is greater than', 'is smaller than'; *c* tendency to use the equals sign in the sense of 'gives as answer' (note Wendy's 'makes five' in the extract), so that they were thrilled when a hint from me uncovered a new set of possibilities of the form '12 = . . .'; *d* confident use of the commutative property of multiplication as well as, most interestingly, of the logical 'and'. Thus, amid laughter, after Stephen wrote '12 > 4 and 12 > 3', Matthew rejoined '12 > 3 and 12 > 4'. I remind you of the fourth reason for talk in chapter one – as a means of flexible assessment relating the other three aspects.

The fact that I can do without the DOING element in this discussion indicates a very confident level of understanding by the children of the various concepts that they bring into play. Note that I follow their lead over the introduction of the different ideas relating 12, 4 and 3. Indeed, some people reading the full transcript have been incredulous that so many distinct statements can be made in such an apparently simple situation. Only over the 'reverse use' of the equals sign was their understanding in need of stretching. This is a very common limitation of meaning, and possible continuation activities might be based around the idea of producing number sentences which are all about the same number. For example, start by writing the sentence

$$3 \times 2 + 9 = 5 \times 3.$$

Other sentences about 15 are then written down in succession – but never including the form 15 itself. These activities could be used to counteract this tendency to ascribe the meaning 'makes' to the equals sign, which unfortunately is likely to be reinforced by the increasing use of the calculator. The whole discussion here revolves around mathematical use of language.

The teacher in the earlier extract 'What's the difference?' has no apparatus. Note how the children, in spite of their familiarity with the subtraction algorithm, can make nothing of the problem until she puts it into a real life context by introducing sums of money. For the DOING element we can sometimes substitute experience that really is familiar to the children – dealing with money, in this case.

But reference to real life experience that they may hardly have met will be of little help in building ideas, and certainly not in testing them. Provision of practical materials is particularly important for the latter; it is regrettable that apparatus tends to disappear from the teaching of mathematics after the infant and lower junior stages. In the next extract we see how a counters game helps two top juniors, first to build ideas and then to test them. It illustrates how the materials are used in different ways at different stages of the activity.

Leapfrogs

Eleri and Emma, two children of average mathematical ability, were playing the game *Leapfrogs* using the board shown in Figure 3.6. There are three yellow and three blue counters (the blue shown shaded). The blue counters can move one place to the right or they can hop over a yellow counter. The yellow counters move in the same way but to the left. The object of the game is to move the counters according to the rules so that you finish with the blue and yellow counters interchanged. (If you are not familiar with this game it would be helpful to try it yourself before reading on.)

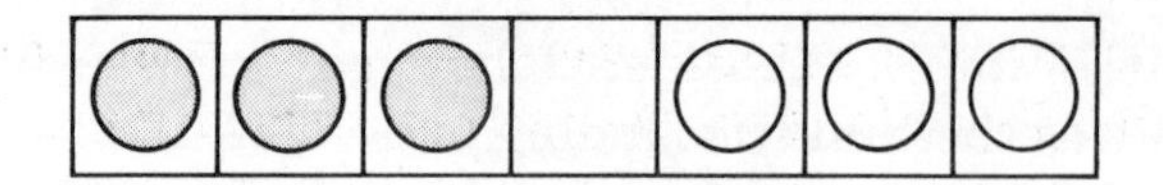

FIGURE 3.6

The children had tried the game before using two blue and two yellow counters, on a board with five squares. After experimenting for a while, the children found a solution to the new version.

Emma: It took fifteen moves – we were able to do it in fifteen moves.
Teacher: Can you show me, please?
Emma: Right – one, two, three, (Speaking as she moved the counters) four, five, six – oh, that's wrong!
Eleri: Try again. Move a yellow counter first, then the blue ...

(After three unsuccessful attempts, the children managed to re-discover their method.)

Teacher: That's good – I found it difficult myself. By the way, this is supposed to be a maths lesson, what's all this got to do with maths?

Eleri: I don't know. It's more fun than doing sums . . .

Emma: Well, you have to do some counting.

Teacher: Oh, yes, the counting. Can you show me again?

(The children repeated the activity, again having difficulty in finding an immediate solution. They continued to count the whole sequence of moves, with each child making suggestions as to which colour counter to move next.)

Teacher: OK. That time you started with yellow and then moved some blue counters. What if we try counting the number of moves we make with each colour?

Eleri: Right – I'll start with yellow. One . . .

Emma: Now blue. Two, three . . .

Eleri: No. One, two – we're counting the number of moves with each colour.

Emma: Right. Yellow – one, two, three – now we've got them all next to each other. Look – blue, yellow, blue, yellow, blue, yellow.

Eleri: Blue – one, two, three. Now it's yellow, blue, yellow, blue, yellow, blue. They've all changed over!

(This time the children managed to complete the changeover successfully.)

Teacher: That was interesting! Shall we write down the number of moves we made with each colour? I'll get some scrap paper. Let's write Y for yellow. Who's going to start?

Eleri: I will. Yellow – one.

Teacher: Blue.

Emma: One, two.

(This talk continued until all the moves had been made. The children had recorded:

Y	B	Y	B	Y	B	Y
1	2	3	3	3	2	1)

Teacher: You did this sort of puzzle with two yellow and two blue counters, didn't you?

Emma: Yes, it took eight moves.

Teacher: Shall we try it again and count in the same way?

(The children agreed enthusiastically and produced the following record:

Y	B	Y	B	Y
1	2	2	2	1)

Eleri: Look, it goes one, two, three, three, three, two, one, up there and one, two, two, two, one down there. There are three threes in this one and three twos in that one.

Emma: Yes, and it goes one, two at each end.

Teacher: Shall we try it for four blue and four yellow counters?

(The children quickly drew a rough board and while counting 'Yellow – one, Blue – one, two, . . . ' etc swiftly obtained the solution:

Y	B	Y	B	Y	B	Y	B	Y
1	2	3	4	4	4	3	2	1)

Eleri: There. It took . . . twenty-four moves.

Emma: We can do it for five counters now. It will be . . .

(She wrote while Eleri chanted:

Y	B	Y	B	Y	B	Y	B	Y	B	Y
1	2	3	4	5	5	5	4	3	2	1

The children adjusted their board to make room for the ten counters and used it to check their solution.)

Emma: For six counters it will be – yellow one, blue two . . .

Eleri: There's no need to check it – I know it's going to work!

It is worth working through this extract in detail, with your own counters. You can see the subtle change that occurs in the children's use of them, first as they struggle to build ideas, much later as a means of testing the exciting patterns that have been built up. Here the teacher has made some very important interventions. Early on they are very typical 'Show me' procedural interventions and only doing and talking are to be observed as the children manipulate the counters according to the rules. Next, the teacher makes a mathematical intervention in the form of a 'helpful question', by saying 'What if we try counting the number of moves we make with each colour?' This improves on their own counting, but lets them keep doing the thinking – note how Eleri reminds Emma of what they are doing a little later.

The teacher judges that it is time to introduce some recording – it is very unlikely that much will be achieved unless this is done, and the children will have nothing but a rather amusing puzzle. The form of recording helps the children to keep track of their thinking, and to see patterns more clearly than is possible by manipulation alone, although repetitious actions can certainly bring a feeling of pattern. Now DOING, TALKING and RECORDING can be seen as the patterns are built up. The teacher reminds them of their previous experience with two counters. It is only at this stage, with two recordings, that the children begin to notice a pattern. So the teacher can intervene again with a probing question about the four counters.

Notice that he does not hint at the pattern, or do the thinking about it for Eleri and Emma. By now they have built up their ideas so that they swiftly get the solution for four counters. Remember that not long before this they had been stumbling experimentally to solve the two counters case! Now they can use their pattern to predict.

Emma says 'We can do it for five counters now. It will be . . .' Notice that the pattern is built up without reference to the counters, which are now used to test the prediction. At the finish they can even generalise to six counters, a tour de force aptly summed up by Eleri's final comment 'There's no need to check it. I know it's going to work!'

In Chapter 1 I mentioned that some teachers were unwilling to admit the possibility of teacher-pupil discussion. This is because the teacher is unlikely to be sharing thoughts and modifying her ideas in the same way that the children are. I think this is probably true of most, if not all, of the teachers in the extracts I have offered. However, these teachers are certainly trying to respond flexibly to the children's ideas, and to get to know as much about their thinking as possible. Whether this type of 'teacher talk' is counted as discussion is probably not of great consequence. My final extract shows a teacher talking with a single pupil. He is trying to help her over a difficulty about decimal place value. Is it discussion?

Trouble with decimal places

The teacher was helping Ann over mistakes she had made in a test. He asked her to say the place values in the money statement '£3.26'. 'Pounds' begins Ann, as he points to the three.

Teacher:	If it were just a number what would they be?
Ann:	Three whole ones.
Teacher:	Three whole ones. What about these? (Pointing at the 2.)
Ann:	Um . . . two tens. (The recording seems quite clear at this point.)
Teacher:	Two tenths?

Ann (hesitantly): Tenths.

(The teacher is unsure whether he heard aright, and 'three pounds twenty-six pence' could justify Ann's 'tens' if she has disregarded, as seems likely, the teacher's earlier 'If this were just a number'. Note the inherent ambiguities in writing money in decimal form! He decides to switch very clearly to a number.)

Teacher: Twenty-seven point five six eight. (Writing at the same time.) Now – what are these?

Ann: Ah – twen – ty –

Teacher: So what does the 'two' stand for? Two what?

Ann: Two .... Don't know!

(The teacher writes '457'.)

Teacher: What about that? What number is that?

Ann: Four hundred and fifty-seven.

Teacher: Now you said it, didn't you? What's the four?

Ann: Four hundreds.

Teacher: Four hundreds – OK. Put 'H' for hundreds, shall we? What about these?

Ann: Fifty . . .

Teacher: So what is the 'five'? Five what?

Ann (after a long hesitation): Fifty tens – five tens.

Teacher: Five tens – OK. And what about those?

Ann: Seven units.

Teacher: Seven units. Now let's go up to this number again – what about that?

Ann: Two hundreds.

Teacher: Look at it carefully again. Read that number out to me – say it.

Ann: Twenty-seven point five six eight.

Teacher: OK – now you said it, didn't you? Say it again.

Ann: Two tens.

Teacher: Two tens – OK, yes –

Ann: Seven units. (Fine! from the teacher.)

Ann: Five hundredths.

Teacher: Why do you say hundredths?

Ann: Because it's past the decimal point.

Teacher: I see – but it's next to the decimal point, isn't it? What comes next after tens, units?

Ann: Hundredths.

Teacher: No, it's not hundredths. Anyway, what about these, what are they?

(He points at the '6' in 27.568.)

Ann: Erm . . . Is it six tenths?

At this stage an interesting state of understanding has been revealed. A little later in the account the teacher writes '3457' and says 'Now if I put the decimal point in, what will the next things be?' Ann replies 'Thousandths' clearly. Her justification is fascinating. 'Because the biggest number there is thousands, so if you go after the decimal point it will be thousandths.' The teacher leads her into the logical consequences by writing.

THOUSANDS HUNDREDS TENS UNITS

and saying 'You write how it goes on'. Ann writes 'thousandths'; he demands 'And the next one?', she writes 'hundredths' and so on, until she has

THOUSANDS HUNDREDS TENS UNITS
Thousandths Hundredths Tenths

Teacher: And the next one?

Ann: Er............................ Don't know! (With great surprise.)

Ann has formed a self-consistent schema about decimal place value which is tested to destruction during this discussion. I shall refer back to this account in Part Three, when I describe Skemp's ideas on schema building and testing in rather more detail. On this occasion the teacher helped Ann to rebuild a more satisfactory schema of place value. He took her sound knowledge of number names (evidenced in the extract) and used the 'ten times' idea, along with lengths as an image, to 'push across' from the whole numbers into decimal places. The talk included an extension of language up to tens of thousands, hundreds of thousands and millions as a preliminary, before reversing direction. Here language seems a powerful guide in extending the *pattern of place value*. The idea of a tenth, or a thousandth *in itself* was not Ann's problem, which had to do with notation and structure. The language for whole numbers which runs in a descent from left to right 'thousands, hundreds, tens . . . ' had led her to the logical-looking conclusion that the decimal values did the same – 'thousandths, hundredths, tenths . . . ' along with a rule about where you started this second descent.

There is historical evidence of difficulty with decimals. Stevin, who introduced the notation in 1585, used a system of priming symbols for his decimals, for example writing the number used in the extract above as

27 5 ① 6 ② 8 ③

It seems to have been some while before it dawned on Stevin and his contemporaries that these clumsy priming symbols were unnecessary, that all that was needed was a 'decimal point marker' to show 'this is the place where decimals begin'. We introduce this extension of place value at the upper primary stage, just as pupils have attained some grasp of its use with whole numbers. It is perhaps no wonder that the National Assessment of Performance Unit found evidence of various kinds of misunderstandings in its tests of eleven year olds. It would seem important not to regard this development as merely a 'straightforward extension of previous ideas'. The increasing use of calculators might help, by providing experience of much larger and smaller numbers, thus reinforcing the idea of a descending/ascending pattern that seems central. So would the continued use of structural apparatus such as the Dienes *Multi-base arithmetic blocks*. The 10 × 10 × 10 block now becomes a unit, with the flats, longs and 1 cm cubes playing appropriate roles, as in Figure 3.7.

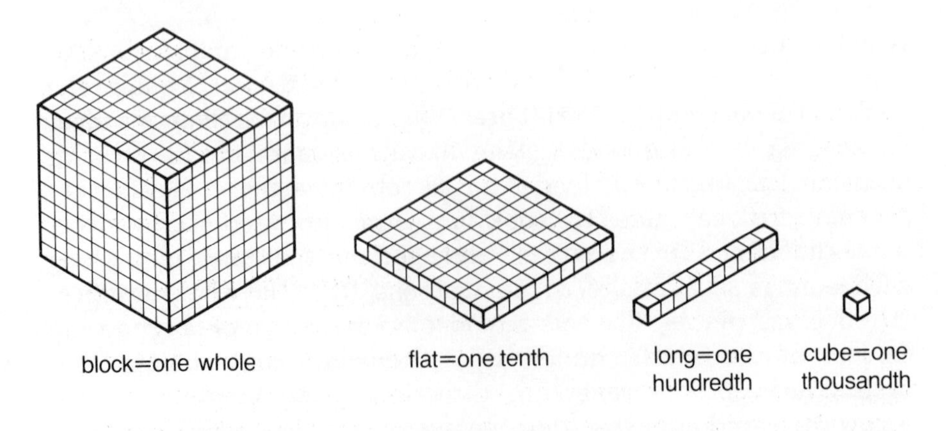

FIGURE 3.7

The teacher's role

In Chapter 1, Figure 1.3, I showed four aspects of the teacher's new role. In this chapter I have shown why I consider it vital that the children are able to build and test their own ideas – why the first aspect, 'planning effective mathematical situations', is essential as a basis of their discussions. It should now also be clear why listening and observing carefully, the second aspect, is an important skill. You will have noticed from several comments that things can often be missed at the time and only picked up later from notes, recordings and reflections. So reflecting and recording could be a useful way of improving one's skills in this respect. The various extracts also contain instances of the teacher 'intervening in ways which keep the children thinking', my third heading in Figure 1.3. Earlier I suggested that there were two aspects to this teacher talk, which I called the teacher's 'mathematical role' and the 'procedural role'. These are developed in Figure 3.8, which lists some of the main kinds of teacher talk under each heading. You will find specific instances in the extracts in the chapter.

Although I have been concentrating on those aspects of the teacher's role in which she or he is working with a group of children, it is important not to lose sight of the fourth aspect shown in Figure 1.3, the organisation of group work so that children are talking together without the teacher. This will require mathematical situations which are effective in exactly the same ways as I have described in this chapter – the presence or absence of the teacher

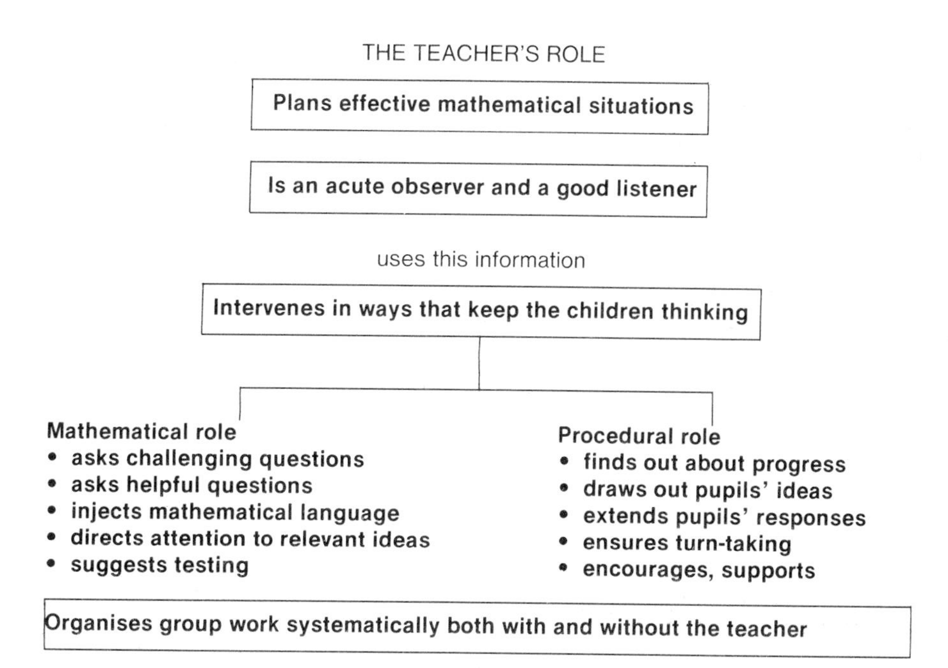

FIGURE 3.8

makes no difference in these respects. The kinds of social skills involved were outlined in Chapter 2, in relation to Figure 2.4. I shall discuss this aspect, the organisation of group activities, more fully in Chapter 7, Part Three.

The whole trend of research in learning mathematics suggests that children build up their schemas of understanding in individual, and rather unpredictable, ways. Recognising this does not mean abandoning a good sequence of activities, but it does mean teaching in a flexible way which makes use of, and is responsive to, their ideas. It is rather easy to dwell on children's misunderstandings and mistakes, an impression which some research reports give, unfortunately. I hope you will agree with me that my extracts are much more cheerful and encouraging. They show that children are creative, can solve problems, devise appropriate methods and test them in an atmosphere of critical but helpful cooperation. Chapter 4 suggests ideas for getting started yourself on working with a small group, and I am sure the results will be at least as good as those I have reported here.

4 Ideas for getting started

Making a start

In this chapter, as I said at the end of the previous one, I shall put forward ideas for getting started on discussion with a small group of children, and for reflecting on the activity. I shall also describe an idea we have found works for us in lessening the initial pressures on the teacher in developing her or his new role. I begin by offering you a step-by-step checklist to help with planning.

1 Choose the group of children that you want to work with, using any criteria for choice that you think appropriate, from your knowledge of the class. I like to work with a mixed group of three, of moderately homogeneous ability, in trying to make a recording, but the extracts in Chapter 3 illustrate groups of various sizes, some of the same and some of mixed ability. The Lakatos group decided that it could not offer any hard and fast criteria – the teacher will select different groups for different purposes at different times. They found no particular difficulty in using groups of boys and girls, but there is various research evidence suggesting that mixed gender groups can work against the interests of girls. One researcher, Tann (1981), even found it hard to form such mixed groups at all! The teacher may need to be alert to dominating or sexist behaviour by boys and be prepared to encourage single sex groupings if necessary. Burton (1986) has edited a very helpful book on this troublesome and discouraging aspect of groupwork. As a general aim I think we would like boys and girls to cooperate in the same groups at times, but to change membership around so that the children get varied experience of working with different people whether chosen by ability, friendship, or other criteria.
2 Settle all the other children thoroughly at tasks that will not require your major attention, so that you can work, as far as possible, without interruptions. If possible, use the idea described below in 'Using another teacher to help' (see page 61).

3 Plan your activity carefully, noting the points I have mentioned before about 'effective mathematical situations'. These are discussed further in Chapter 6, but particularly bear in mind our icon of DOING, TALKING and RECORDING (Figure 1.2) and the associated idea that your children should be able to BUILD AND TEST IDEAS (Figure 2.4). You will probably find that it is best to begin with a practically-based activity.

4 Arrange the group so that they sit facing each other (*not* in a row) perhaps on a carpeted area of the floor or around a table. This attention to seating may seem a minor matter, but it has been found to be of vital importance in creating a cooperative group atmosphere in which no one feels left out.

5 Sit with your group and make sure that they *know* you want them to talk.

6 Talk to them first about the activity, about any apparatus and how it is to be used. It is usually best to have a central stock of materials in the middle of the group, from which they can help themselves as required – for example Unifix cubes or Cuisenaire rods. This can save a lot of argument and encourages cooperation. In the cube game described in Chapter 3, for instance, we had a central pile of blue and yellow cubes, and the shapes were made on a single large sheet of squared paper. Similarly, if a group is making written recordings, give them a single large sheet such as sugar paper – this will be another effective way of getting the children to share their thinking.

7 During the activity, OBSERVE, LISTEN and RECORD. You will learn a lot, about some or all of the following aspects:

- what the children are thinking;
- how much progress they are making or have made;
- how they are influencing one another;
- which children need help;
- which children give help;
- their articulation of ideas;
- their social skills;
- leadership, dominance or withdrawal;
- their dexterity skills in handling apparatus;
- their imagination;
- their perseverance;
- what they need to do for their next activity.

8 Sometimes you may find it useful to draw a diagram to record patterns of activity within the group. Use coloured arrows to show the direction of different types of behaviour. For example, Figure 4.1 shows the predominant pattern of 'helping' (thick

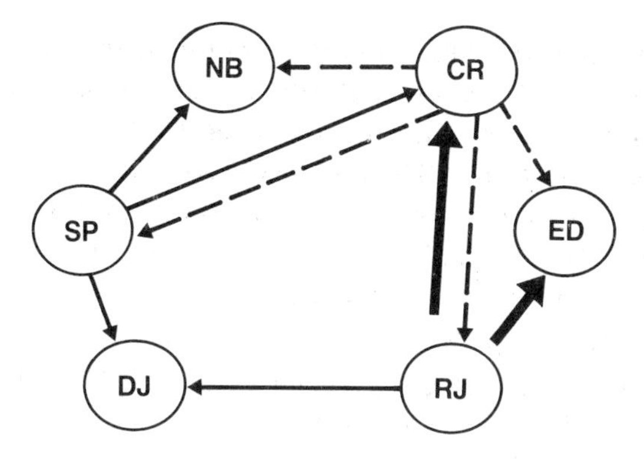

FIGURE 4.1

arrows), 'talking' (thin arrows) and 'watching' (broken arrows) during an interlude with a group of six children.

Making records of the activity

You will want to make written notes of things that were said, either during or immediately after the activity. I shall offer some guidelines for reflecting later on. Making these notes in writing is probably the most comfortable way at the beginning. Making a cassette recording of part or all of the discussion can be very useful, of course – such recordings were the basis of many of my extracts in Chapter 3. But the teachers involved had all used recordings previously, as part of an in-service course, and many of the children were accustomed to such an approach. So I would recommend caution over using tape recordings at the outset – only do so if you feel very confident. The main purpose is to concentrate on the aspects of your new role, after all! Children tend to be tense and nervous or to become distracted by the presence of a recording machine. If you are nervous as well, conversation will be stilted or even dry up entirely! This would destroy the whole point of the activity for you. Most such recordings are only likely to be useful to the teacher concerned – it is very difficult for other listeners to identify individuals and to pick out what is said – although it will have been perfectly clear to the participants. So be prepared to be disappointed about the quality of most of your attempts, and bear in mind that this difficulty restricts the usefulness of such records at staff or in-service meetings. I have overcome this problem by transcribing and duplicating short transcripts of important episodes in the recording, lasting two or three minutes. This does help listeners to follow a recording of reasonable quality. The problems of making recordings of good quality should not be underestimated – even the professionals, with an array of equipment and assistance, can run into trouble.

Conducting discussion with a group – a summary

1 Be a good listener and an acute observer.
2 Use this information to judge when and how to intervene – remembering that it might be better to hold back.
3 Keep your mathematical interventions to a minimum – let your planned activity work for you and the children in helping them to build and test ideas. If you do intervene, try to use questions rather than explanations or hints, in order to keep the children thinking.
4 Don't try to push the discussion in the way you think it should go – always try to use the children's ideas, not your own!
5 Forget your ideal responses – listen to what the children actually say.
6 Stop thinking it is your job to comment on everything that is said. If necessary, use procedural interventions, such as I gave in my extracts, to bring in children to comment, or to help one another in cases of difficulty.
7 You will often diagnose misunderstanding by listening and observing. Get other children to help, as above, or file children mentally for some individual attention later, on the basis of what you have learnt about their thinking.
8 A probing question *at the right moment* will often produce better work and deeper thinking from the children.
9 If things start going wrong, and you find that you are doing most or all of the talking, think afterwards how you could re-plan the mathematical situation so that it helps the childern better. You might need *a* more carefully-prepared materials for them to use; *b* better-structured questions to start them off; *c* more opportunities for them to think, compare ideas or help one another.
10 If any misbehaviour occurs, deal with it much as you normally would, from your knowledge of the children, but if possible in an 'aside', so as to minimise disturbance to the working atmosphere you have created – rather as rugby football

referees give offenders a quiet early warning, rather than a public confrontation.

In Figure 4.2, I have included some typical questions that you will certainly find useful in your new role.

Can you find?

Show me . . .

Let's try . . .

Tell us . . .

What do you think . . .?

Can you tell us what you mean?

How did you get . . .?

Do you think so . . .?

What if . . .?

Will it work for . . .?

FIGURE 4.2

Using another teacher to help

At in-service meetings, as I said in Chapter 1, I am constantly asked 'But what are the rest of the children doing?' In the long run, of course, teachers must organise their class activities so that, as I said in point 2 of my first checklist, they can work as far as possible without interruptions. This will mean planning contact time with the different groups carefully. But at the beginning, it will be very helpful if you can get the help of another teacher, to 'run the class' while you concentrate on your group. This idea has been tried in various circumstances and we find it works very well. People on whom you might draw for such support are the headteacher, the mathematics coordinator in your school, or a visiting educational support grant (ESG) teacher.* In the first two cases, this might be as part of a school-based in-service programme. Such a programme might be part of the school's staff development plan and related to appraisal. Another possibility is that the mathematical activities of the school could be so organised that the headteacher, or the

Educational Support Grant, or ESG, teachers were outstanding teachers who were seconded from normal teaching duties by local authorities, with the aid of Government funds, during the three-year period 1985–88. They visited schools to help in the development of teaching methods in science and mathematics, and assist with in-service programmes.

mathematics coordinator, could help different members of staff from time to time in this way. This approach might be particularly helpful to probationary teachers joining the staff. An ESG teacher needs to plan a careful programme of cooperation with schools in order to offer such assistance. One ESG teacher made the following comments about her own programme:

> This idea has worked well for me. I developed the following strategy from a comment a teacher made to me about how difficult it was to give an idea a fair trial in a full and busy classroom. I made an offer, 'What if I take your class and you take a group?'. So it was arranged – I visited the teacher weekly, over a whole term, and released her for an hour, during which she always managed to repeat her trial with two different groups. On the first occasion I simply assumed an invigilating role and got to know the children. On subsequent visits, I planned some innovation of my own (introducing calculators in this instance) with at least one group at a time, while supervising the remaining two groups in the class as they worked on other topics or tasks. During the one hour session we would rotate the groups (as shown in Figure 4.3). Thus each group would spend some time with a teacher every week, and twice every fortnight, during our sessions.

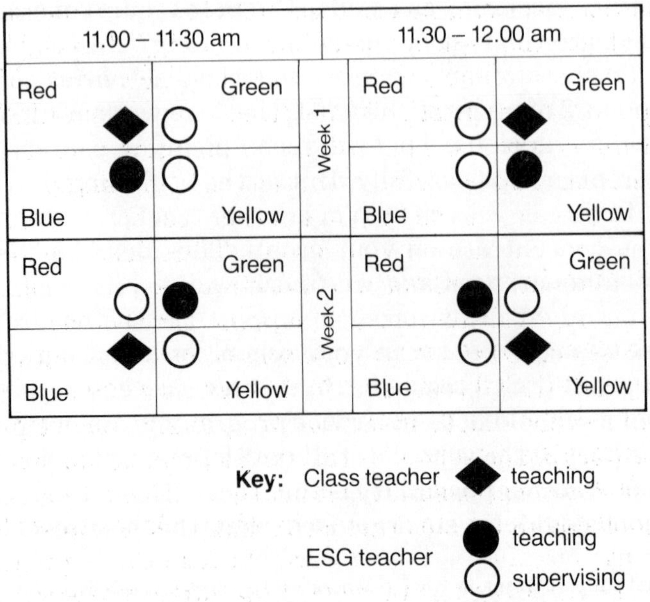

	Group	11.00 – 11.30 am	11.30 – 12.00 am
Week 1	Red	◆	○
	Green	○	◆
	Blue	●	○
	Yellow	○	●
Week 2	Red	○	●
	Green	●	○
	Blue	◆	○
	Yellow	○	◆

Key: Class teacher ◆ teaching

ESG teacher ● teaching ○ supervising

FIGURE 4.3

Note the pleasing symmetry displayed in this rotation diagram (Figure 4.3). It is not, of course, the only solution, but it shows how to work with four groups (here, called Red, Blue, Green and Yellow). Figure 4.4 shows a similar rotation that was used by the same teacher, with six groups split between the ESG and class teacher. Each teaches one group and supervises the other two. The latter need not be doing mathematics - it could be activities suited to the age group such as construction, model-making or shopping.

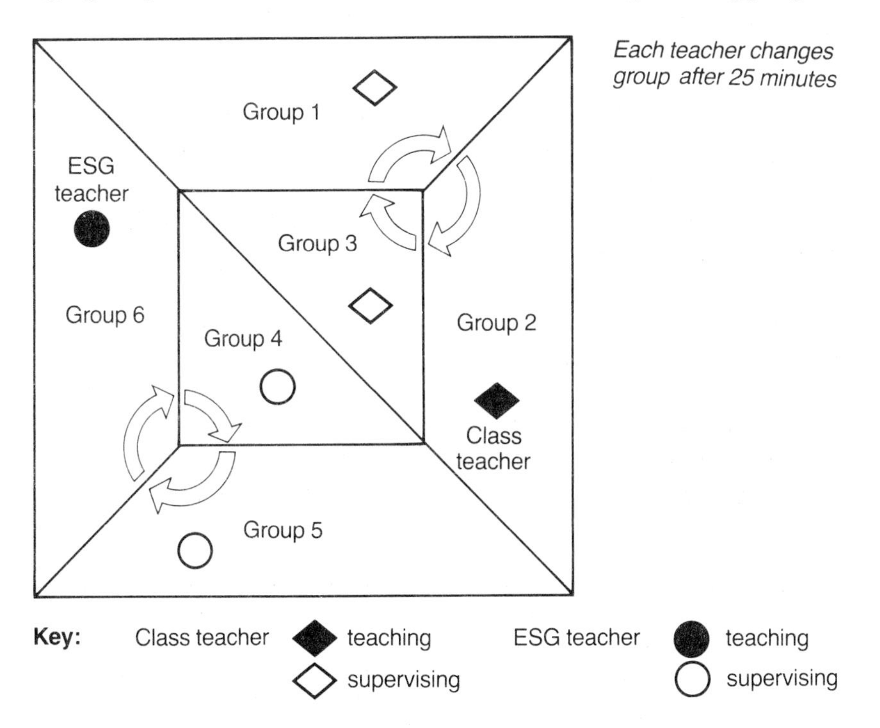

FIGURE 4.4

The ESG teacher continued as follows

> These plans proved very successful in terms of practical help to teachers and were extended to include a second teacher in the school. This made a whole morning or afternoon possible, in one school, each week.
>
> I made some general observations:
>
> 1 It is important to make time for discussion with the teachers concerned so that you are aware of each other's progress.
>
> 2 It is encouraging to note how often teachers use the intervals between my visits to do extra work.
>
> 3 Teachers' confidence is improved, since they have the

opportunity to concentrate on refining techniques with small groups.

4 It is important for the visiting teacher to be ready to (i) provide ideas for further development, (ii) modify his or her own ideas when trial with children reveals things that do not work.

5 Teachers are invariably interested in what you are doing mathematically with their class at these times, so do not waste the opportunity.

6 Teachers do not feel that things are being 'handed down', they feel more involved.

7 Teachers' own experiences in the activities can give valuable information.

8 Involvement with shared teaching of the same children aids understanding of particular problems, for example special needs.

9 BUT, the main advantage remains - uninterrupted teacher time, in the initial stages of trying out ideas and refining discussion skills.

Reflecting and recording

In these early stages I strongly suggest that some further work reflecting on the recordings of the group activities be done as part of a school-based in-service programme. As I have already said, the act of attempting to make some form of recording, and the study of it later, does seem to sharpen one's skills of listening and observing, as well as developing insight into possibilities. It may sound to experienced teachers rather like teaching practice all over again, but the benefits are there! I have certainly gained in these ways myself, and become (I think!) a better teacher. Again I can offer some guidelines that have worked with groups of experienced staff, and also show some of the results of these reflections.

The teachers were asked to study their original notes and then to write further comments in the light of the following pointers:

1 Were you able to note and reconstruct the main statements made by the children?
2 Can you reconstruct and describe the general 'flow' of the activity?
3 What were the 'significant moments', if any, during the activity?
4 What sorts of impressions did you record concerning behaviour,

types of understanding or misunderstanding, and kinds of group interaction displayed (eg cooperation, withdrawal, dominance)?

5 Can you analyse your own part in the activity, particularly in the light of the suggestions that have been made about the teacher's role in discussion work?
6 To what extent do you think you followed the suggestions that have been made about this role? (Be as honest about this as possible!)
7 Can you find times where you were really 'teaching conventionally' rather than 'conducting discussion'?

Results from this activity have been most encouraging, and show that teachers can soon get to grips with the ideas that have been put forward, although a further 'developmental period' is likely to take place. Here are some of their accounts and comments.

1, 2 and 3

'My friend painted his front door and then bought three numbers to fix to the door: 1, 2 and 3. What was the number of his house?'

This idea came from '*Pointers*' (Cambridge University Press). I thought I would try it with a group of eight year olds who had been working on place value using base 10 apparatus. Beforehand I bought some large plastic numbers from a 'car spares' shop, stuck them on to cards and covered them with clear film.

The problem was quite straightforward, so I thought I could make it more interesting by putting it in a story. I elaborated quite a lot - making my friend out to be rather larger than life and very absent-minded - unscrewing the numbers and forgetting the order after painting his door bright orange!

At first the children played around with the three numbers and it was a useful reminder to get them to say 'one hundred and twenty-three', 'three hundred and twelve' and so on. This helped to reinforce the concept that a particular digit can represent a number of units, tens and hundreds according to whether it comes first, second or third reading from right to left. Also it helped to assess the children's abilities;

a to identify separately which digits represent units, tens and hundreds;
b to read aloud two and three digit numerals;
c to match these with physical representations of the numbers.

(Reminder from Richard Skemp - 'This is a brilliantly simple notation whose brilliance is easily overlooked because it is so simple.')

At this point we were sidetracked into talking about our own door numbers and the different ways they can be arranged, using the plastic numerals as a means of recording this. (On reflection, next time I think I would use this as my starting point.) I didn't want to go into telephone numbers at this stage – Mark brought these into the discussion – so I returned to my friend and his problem. There was a great deal of discussion by the children which I was unable to

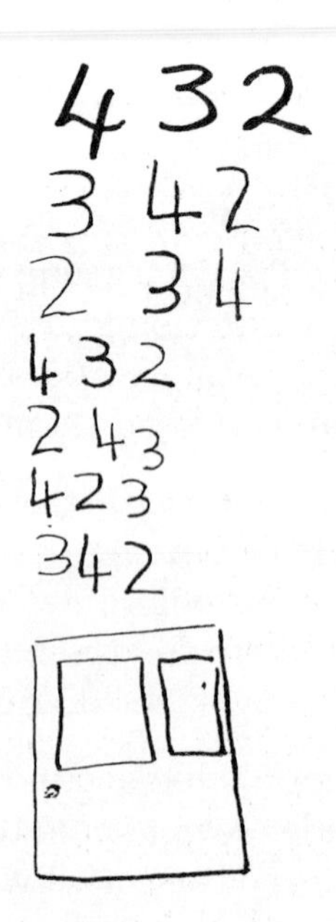

FIGURE 4.5

record in detail, so I have tried to think of what happened using the framework we have been given for reflecting (see page 64).

Replies to the points in the checklist in order:

1 I can recall that the children's thinking at this stage seemed to be in parallel, they all tried random methods to get as many different numbers as they could. (See Figure 4.5.)
2 The general flow of the activity was that after a few minutes they began to note the *number of possibilities* they had recorded, saying things like 'We've got four now, is there another way?'
3 The 'significant moments' were when the children could not get any more than six different numbers each and were given paper for recording what they had found. I avoided using nought or any digit twice - this may be a useful continuation activity. I was also careful to avoid giving any hint that six was important (from this first joint effort) and encouraged the children to keep on looking until they were satisfied and Louise announced '*There are no more ways*'.
4 In this group there were no noticeable features of withdrawal or dominance, in fact I was surprised by the quality of the group interaction and cooperation. (They are well used to working together.) I think the discussion about their own door numbers helped their understanding and it served as a 'simple case' as well, because all but one of the children had houses with two-digit numbers.
5 I was very aware of my mathematical and procedural role and knowing about it helped to identify the interventions I made.
6 Honestly! I did follow the suggestions made.
7 Honestly? No - I can't find any times when I was 'teaching conventionally' rather than 'conducting discussion'. The early emphasis on the place value language I considered to be appropriate in 'extending the children's use of language'.

At this point the children began to feel that they had reached the end so I asked them to try to write in a sentence what they had discovered. I hadn't hoped to get any further than this but was interested in Ryan's conclusion, that he was sure he couldn't get any more because he had 'done a test'.

> What do you mean, Ryan?
> *Well*, he explained, *look at the start number and twist the other numbers.* (He couldn't write this himself, but could talk about it.)

Everyone listened carefully, then tried it out with their own numbers – see Figure 4.6 for what they produced. They all agreed that it was a good way of making sure that there were no more numbers. I was delighted at this outcome and felt sure the activity could lead on to even more interesting work later.

As we were tidying up and chatting about what we had done I asked, almost without thinking, 'How could my friend check he had the right order?' I had a whole lot of ideas! Here are some of them:

He could ask his wife.
Leave off the numbers and wait for the postman to come.
He should have written it down first!
Look to see if there was a shadow from before he had painted.
Look at the house next door – next door is always one away.
No it's not – sometimes it's two.
It's best to check both doors and see what number is in the middle.
Well, look on his hospital card – Or his membership card.
Why? – Because your address is always there.
See how the screws fit . . . look where the screws are on the numbers and fit them on the holes on the door.

My final example comes from a teacher who gave a fine account of her small group investigation at a staff in-service meeting, and generated much discussion. Her record is described below.

At the shop

The group consisted of five infants aged between six and seven years: Simon, Angharad, Takuto, Becky, Lindsey.

Four children were each given a card-purse containing 15 pence in assorted denominations. The shopkeeper, Lindsey, was given a tray of assorted coins. The goods consisted of 18 small items (toys, sweets, fruit, jewellery, perfume and biscuit bars) each labelled with low-denomination coin prices.

Teacher: You can all go and spend your money now and buy what you like.

(At this point, they proceeded to select what they wanted and were remarkably polite to the shopkeeper! They chatted away, noting prices carefully and considered what they would and could buy very sensibly.)

Angharad: I've bought too much and I've still got two pence

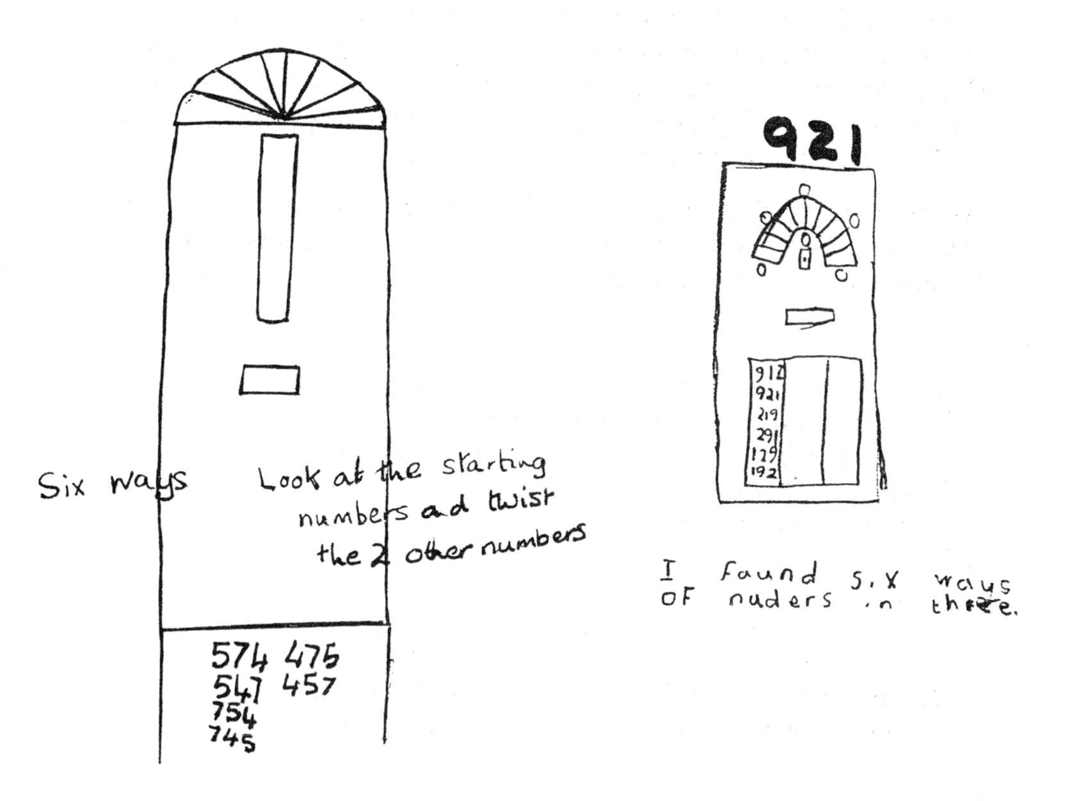
Six ways
Look at the starting numbers and twist the 2 other numbers
574 475
547 457
754
745
921
912
921
219
291
129
192
I found six ways of nuders in three.

736
673
637
736
7 6 3
367
376
I had 736 and we found out we had six

FIGURE 4.6

left. Two pence is too much for that badge though.

Teacher: You can't give things back, but Lindsey can give you change.

(The youngest child, Becky, bought a fan for 4p. She seemed either reluctant to buy more or was unaware that she still had 11p left. There was plentiful discussion but mainly about their purchases, especially the edible variety. The children chatted, giggled and generally enjoyed their encounter with money problems.)

Angharad (echoed by Simon): I've spent all my money now!

Teacher: How many things have you bought?

Simon: I've got seven things and I'm still thinking about eating them.

Angharad: I've only got five things and I could eat two of these.

Simon: I bought cheaper things.

Angharad (counting again): One, two, three, four, five . . . six.

Takuto: What is 'cheaper things'?

Simon: It means less money, because if you have less money . . . Suppose if somebody had one pound and if you had fifty p and you wanted to buy the same number of things you just have to buy cheaper things.

Angharad: It means you can buy more for less money.

The account of how the children explained 'cheaper things' to Takuto, a Japanese boy, was particularly interesting to the participants in the follow-up staff discussion. Would it be more helpful to Takuto than his teacher explaining it, do you think? Another intriguing point to note is how accurately the young children seem able to 'role play' the real-life situation of being at the shop – for example in their politeness to the 'shopkeeper' or Angharad's comment about the price of the badge!

School-based in-service training

Much in-service training will be needed to help primary teachers to carry out the recommendations of the Cockcroft Report. In 1985 a major project was launched, under the leadership of Hilary Shuard, then President of the Mathematical Association, to develop local initiatives on which such a national programme might be based. This project, *Primary Initiatives in Mathematics Education*, (PrIME for short), aims to produce INSET packages for various aspects of the curriculum, including the use of calculators and microcomputers, investigations and problem solving, and - particularly relevant to our own point of view - the need to bring about a better balance of teaching styles in mathematics. Such programmes have many advantages if they are school-based rather than bringing together teachers from different schools at another centre (although these programmes have their own merits, such as the opportunity to compare ideas and experiences). The advantages of the school-based approach have been described by one head teacher in the following terms:

1 *Promotes internal school development* Traditional INSET courses are of great value to the individual but rarely have much impact on the school. The impact is confined mainly to the member of staff who attends the course. Even when this teacher writes a paper for the other staff the impact is lost or lessened for the others as each member will take something different away from it.
2 *The needs of the school are catered for* Each school has needs which are peculiar to itself. Courses based at the school can be organised to take these needs into account and to show how they can be met and overcome. These needs can include: constraints of the school building; lack of suitable resources; lack of staff expertise (ie no-one on the staff with a specialist mathematics qualification.)
3 *There is more scope for discussion* Generally, people are more likely to discuss their problems or difficulties in small groups and need to feel secure before they do so. This security can be achieved in a small staff group on their 'home ground'.
4 *Known pupils are used* The observation of known pupils at work on newly-introduced ideas dispels feelings of a situation being set up 'especially for the purpose'.

The ideas and general approach which I have put forward form the basis of such a school-based programme, as will be apparent from the experiences on which I have drawn. In the first three chapters I put forward the basic introductory material. In this chapter I have outlined the key ideas for the actual programme, which could take perhaps a school term and involve a series of staff meetings. Figure 4.7 shows the general form this programme might take.

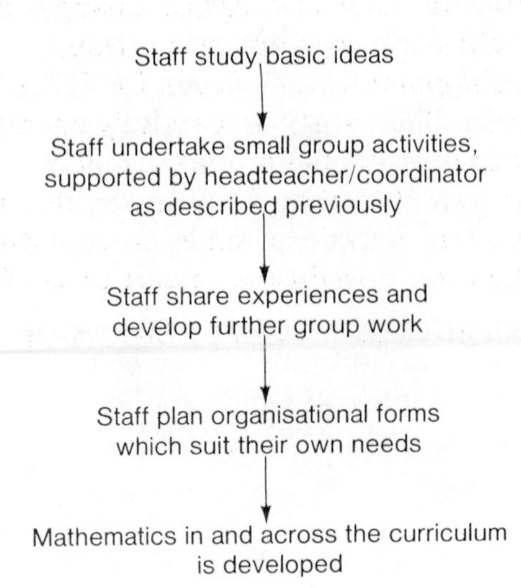

FIGURE 4.7

The later stages of such a programme of staff development might include enlisting parental cooperation and the preparation of suitable materials for this. This would be particularly important when the school is planning major changes, such as basing the number work around the use of calculators. Examples of such possibilities are offered in Part Three.

The need for such a process of staff development might be identified as a result of a school self-review undertaken through the GRIDS method. GRIDS is the acronym for *Guidelines for Review and Internal Development in Schools*, a project funded, like PrIME, by the Schools Curriculum Development Committee. It sets up guidelines, as the title suggests, for the process of conducting a school self review. Five stages are envisaged in the process, as follows:

1 *Getting started*: when the organisation, resource implications and choice of school coordinator are considered.
2 *Initial review*: staff are invited to express their views of strengths and weaknesses via an anonymous questionnaire, so as to identify areas which require a specific review.
3 *Specific reviews*: Each such review (usually from one to three) is analysed by a review team led by a coordinator. This would obviously be the mathematics coordinator in our case. This team

analyses strengths and weaknesses of present practices and makes recommendations for changes. Changes which are agreed by staff as a whole are then put into action.

4 *Action for development*: staff needs for INSET are identified and set in train. This stage is carefully monitored to gather evidence that improvements are taking place.
5 *Overview and restart*: Comparisons are made between past and new practices, and decisions made on retention, rejection or further modification, possibly by restarting the GRIDS process.

The guidelines booklet was written by McMahon *et al*, (1984).

5 Talk using microcomputers and calculators

The way in which microcomputers have come to form a part of the primary school curriculum is one of the success stories of British education. They first appeared in the early 1980s, but there was a dramatic upsurge in their use between early 1983 and mid-1984. This resulted from a major government initiative in providing a 50% grant to schools for the purchase of their first microsystem. The vast majority of local education authorities purchased the BBC microcomputer and its associated monitor and tape recorder, although its memory (32K) was limited. There have been subsequent improvements, so that micros supplied in 1986 had a vastly improved memory of 128K, and many schools have been able to purchase disc systems to replace the clumsy cassette recorders which caused many difficulties in the early days. The outcome of this major capital investment in an excellent, if elderly, design, is twofold. On the one hand, most of the educational software for the primary school has been written with the BBC machine in mind. On the other hand, most of the curriculum development and in-service training has similarly used the same equipment. In spite of the inevitable limitations, much of this work is praiseworthy and often exciting.

In the case of mathematics, at least, the primary school developments compare very favourably with those at the secondary stage. It has proved far easier for primary schools to make use of their one or two microcomputers than for secondary schools, which with much larger numbers and subject-based forms of organisation, have often found it well-nigh impossible to bring their limited numbers of machines into the teaching of mathematics. It is not unusual for children who have had weekly experience of micro activities in their primary schools, to find nothing at all of this kind in their secondary school. I have talked with numbers of children who have felt great disappointment at this abrupt change in teaching methods. Increasing numbers of children will have computers at home, of course – often a different computer from the one they use in school. Straker, to whose ideas I shall refer later, found in various small surveys that about forty per cent of eight to twelve year olds had a micro at home, which they mostly used either to programme in Basic or to play arcade games. Meanwhile, at work, their parents have switched almost entirely to the new technology. Fitzgerald (1985) found the use of the calculator almost ubiquitous

> It has become very rare indeed to find an employee performing multiplication, division and percentage calculations, except sometimes the simpler kinds, using written methods. It is a little less rare for subtraction and addition, although for the latter a calculator will almost invariably be used if more than a handful of figures have to be totalled.

Elsewhere, the computer plays an ever-increasing role – for example it has replaced the drawing board and T-square in the design of cars, highways, buildings and aeroplanes. There are obviously major implications for the mathematics curriculum in the primary school.

In order to help with the teacher training needs, the government set up the *Microelectronics Education Programme* (MEP); within this was a specialist Primary Project under the direction of Anita Straker, which operated until 1986. This advised local education authorities on needs and produced free training materials and a free set of programs, the *Micro Primer Pack*, which all primary schools received with their first microcomputer. Although many other materials have since been developed, all primary teachers will be acquainted with the thirty-three programs in this first kit and will recognise some of them in my subsequent accounts. As Pearson (1986) observes, in a very helpful booklet, 'In one sense, finding the money to place this equipment in schools was the easy part of the development. Deciding on the most appropriate uses for it and getting that information across to teachers has been more difficult.'

It is not my purpose to examine these aspects in any detail, for much more expert accounts have already been given elsewhere. Shuard's presidential address to the Mathematical Association (1986b) covers many of the important findings and puts the case for the 'Calculator Aware Number Curriculum' which forms part of her PrIME project. It is certain that calculators and microcomputers will play an ever-increasing role in primary mathematics, and bring about radical changes. It is the part played by talk and discussion in this with which I shall be concerned in this chapter. Readers will need to bear in mind that my accounts will, because of the pace of technological changes, be out-dated to some extent almost as they are written – for example, one of my extracts refers to the use of cassette recorders for program storage, a method that is rapidly vanishing.

By a happy chance, the conditions necessary for talk and discussion that I have set out in preceding chapters have been virtually forced on teachers in order for them to make effective use of their microcomputers. About this Pearson, *op cit*, writes:

> Though the management issues of computer assisted learning will not be discussed here in any depth a brief comment on pupil access to the technology would seem appropriate. Programs can be used with a whole class, with small groups of children, or by an individual; whole class and individual use are not frequent. For most children, most of the time, a group of three or four appears to offer good opportunities for discussion, evaluation, problem solving, consensus decision making and the development of social skills.

What more do we need in order to bring in, if we wish, the approaches based on talk and discussion?

Group work and the micro

There seems no reason to suppose that the idea of an 'effective mathematical situation' as discussed in Chapters 2 and 3, should not be applied to our micro programs. Some will work well with a small group, others will only allow one pupil actually to do the mathematics – whether others happen to be present or not. In the latter case, this will probably be the person at the keyboard. You may notice this in the extracts which I present. Some hard thinking about the relationship of my DOING, TALKING and RECORDING icon to computer programs will also be needed. It is likely, as I shall suggest later, that many programs may need to be set carefully into other materials written by the teacher, so as to form an integral part of an activity, rather than seen as 'standing in their own right'.

Mathematics 5–16 follows the Cockcroft Report's classification of the learning of mathematics into five areas:

A Facts
B Skills

C Conceptual structures
D General strategies
E Personal qualities

Under Objective 9 in the same work we read that the microcomputer can be used as a teaching aid, a learning resource, or as a tool for pupils in doing a mathematical task. So far many of the primary mathematics programs are directed at aspect B, the practice of skills. About this the Cockcroft Report says 'They need not only to be understood and embedded in the conceptual structure but also to be brought up to the level of immediate recall or fluency of performance by regular practice.' Such programs support existing aspects of the curriculum, but, continuing a view put forward by Straker, do not enhance it or provide new opportunities for change. We would like programs which support or enhance all five areas listed above. This would certainly happen if micro programs could support group work and discussion, for instance. In order to examine this possibility, I shall attempt to look at the microcomputer from a different point of view in the next section - by thinking of it as a kind of 'extra member of the group'. One can observe this tendency to treat the computer as a kind of 'person' in statements like 'Tell the computer to . . .' or 'The computer says that . . .' Some programs invite pupils to type in their name at the beginning of a session, and use this information in subsequent feedback - 'Well done, Philip!' or 'Try again, Brenda!' and the like.

The micro as a 'group member'

With a program loaded and running, the children can communicate with the computer and it can communicate with them. The quality of this communication depends on the design of the program and on the kinds of responses which the children make to it. In this sense, the situation is not unlike group interaction, where what happens depends on what the participants are willing and able to do together. The person - computer communication is not talk (that possibility is still a long way off, in my opinion); it requires us to think of another element in the interaction, as well as the children and the teacher, as in Figure 5.1. The micro may be a teaching aid, but it is an exceptional one in this communicative sense.

The microcomputer program constructs a 'microworld' with which the children can communicate and which communicates with them. This 'microworld' is a *separate reality* with its own laws and rules. Some microworlds allow the children to make their own laws,

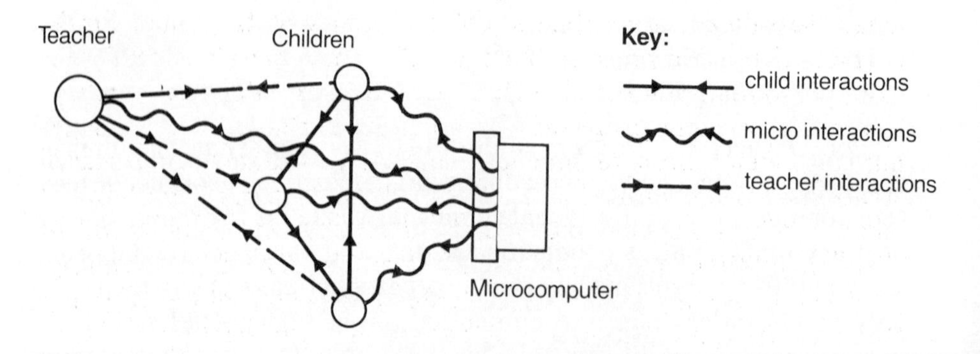

FIGURE 5.1

operations and structures – the well-known programming language Logo being probably the most notable example. Others afford opportunities for decision-making and problem solving – for example the popular simulation *Granny's Garden*. 'Practice' type programs, such as *Trains*, provide straightforward microworlds, albeit in a motivating format of sounds and colours, where the child-micro communication is of a lower quality. The microworld seems to provide the DOING of my original icon (Figure 1.2), and maybe some of the RECORDING as well. I shall therefore modify Figure 1.2 to place the microworld at the centre of the interplay (Figure 5.2) whenever the mathematical situation includes children working with a program.

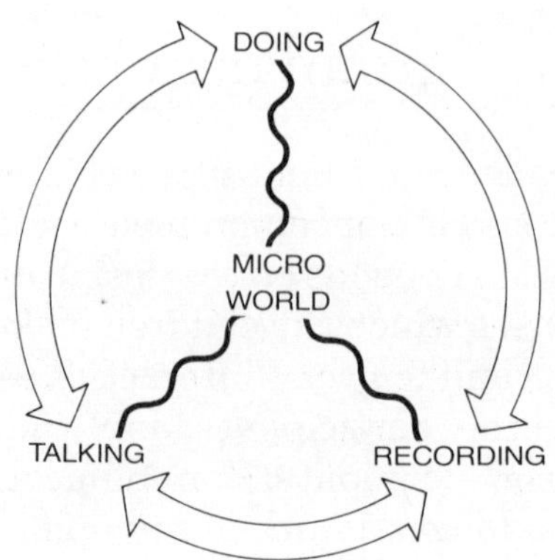

FIGURE 5.2

This idea of a microworld fits in beautifully with Skemp's BUILDING and TESTING. It is clear that microworlds can vary enormously in the extent to which they help children to build their ideas, just like the mathematical situations of the usual kinds. Some really offer no such help at all - they merely TEST - for example, programs which test knowledge of tables or spelling. There are various programs based around the idea of treasure hunting, which involve concepts such as coordinates and search strategies. Microworlds like these clearly do help in the building of ideas, but it seems very likely that the teacher will have to embed them in a wider set of materials which develop understanding. Microworlds of any kind offer the tremendous advantage of giving feedback to the children about their communications with that world. This means that the children have a means of testing their ideas, essential in an effective mathematical situation, *without bringing in the teacher.* But where microworlds give little or no help in building, the teacher will have a considerable task in teaching, or diagnosing and helping where there is misunderstanding. We now have a way of evaluating the usefulness of particular programs, deciding where they fit into our scheme, and identifying what other materials and activities might be needed in addition to the program itself. This is by asking the extent to which they help our children to build, and then to test, the ideas involved in the program (Figure 5.3).

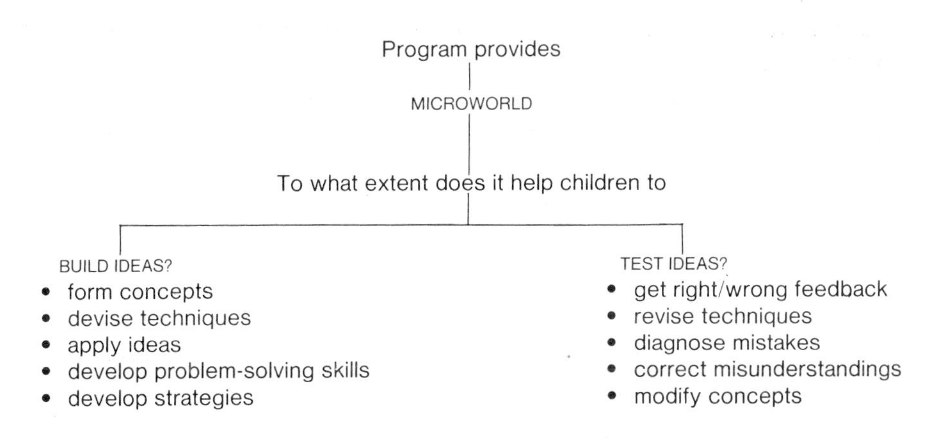

FIGURE 5.3

So far I have not considered the idea that the microworld is interacting with a group in any detail, although this was suggested in my first diagram, Figure 5.1. There are a number of possibilities, which are suggested in my next Figure, 5.4. First, the microworld may interact just with one member of the group, with the others remaining as relatively passive onlookers, as in 5.4 (*i*). Second, all members may interact with the micro, but little (or not at all) with

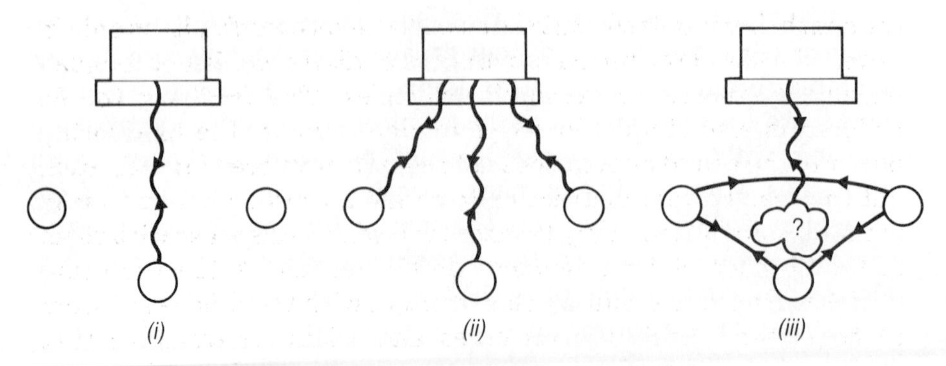

FIGURE 5.4

each other, as in 5.4 (*ii*). A situation that we would very much like to create, I think, is that shown in 5.4 (*iii*), which is intended to suggest that the microworld *presents a problem or idea to the group, which they consider together*. All these possibilities remind us of things that can happen in the ordinary group situation. For example the 'whizzkid' who virtually rules over communication with the computer is like someone who dominates a group. There is a clear tendency on the part of boys to do this in a mixed group, so some firm intervention might be needed on occasions. In the same way the teacher might get so involved with a program that she, or more likely he, loses touch with the group of children. Another possibility is when the teacher competes with the microworld in holding the attention of the pupils - you may well have experienced this. There is also the question of seating - of vital importance, as I said in Chapter 4. How should we seat a group of several children so that they can all have access to the keyboard, and yet are not in a row?

I shall now attempt to apply the ideas I have developed in this section to a number of accounts of children using microcomputers.

Children using 'TRAINS'

Trains was one of the programs distributed to schools in the *Micro Primer Pack* referred to earlier. It provides practice in the skills of addition and subtraction by setting sums at varying levels of difficulty, in a motivating context of building up pictures of trains by adding coaches to an engine. The microworld is lively enough, but somewhat limited. The level of difficulty can be selected by the teacher or the children, and there are optional sound effects as a coach clatters along to join the train, following a correct answer to the sum being inputted. But the so-called level of difficulty is only in terms of a simple criterion of numbers of digits and is not founded on any diagnostic or research principles. The feedback to the children is also simple, in terms of right/wrong. The motivating power of the sound effects does not seem to have been investigated, but they can be very distracting to children working at other tasks in the classroom and many teachers will wish to dispense with them in the absence of firm evidence on the matter. A teacher using *Trains* will need to maintain close contact with the children in order to help them tackle the mistakes and misunderstandings that occur. This could be done by using the approaches in the following extracts, but the program keeps no detailed record. The future of a microworld of this type is clearly in some doubt because of the introduction of calculator-based methods.

The teacher in the following extract made recordings of pairs of her children using *Trains* gaining useful feedback about their thinking. About her organisation she writes as follows:

> My class uses the computer two days a week, organised in ability groups to consolidate their mathematical concepts. When working in pairs the children become far more actively involved and cooperate enthusiastically on the task. Occasionally I pair a child who has experienced difficulties with a more capable child, to help him or her to gain confidence. During the sessions, groups not directly using the computer engage in activities less demanding of my time – this allows me to interact with the computer group properly. But from time to time I allow the children to form their own social groups and pairs, to work on programs of their choice for enjoyment, within the same form of overall organisation.

A low ability group – Gareth, Mark, Leanne, Steven, Craig, Kelly (three pairs)

The group work as three pairs on the problems set by the computer, using individual number lines for counting on and back, as well as referring to the class number line if necessary. Some typical comments from the children:

Gareth: Eight take away one – that's one less than eight so it's seven.
Mark: Seven take away seven equals nought. Nothing left.
Leanne: Seven take away four – I can count back four on the number line.

Steven: Eight add seven – that's eight, nine, ten, eleven, twelve, thirteen, fourteen, fifteen. (He counts on aloud from eight, using his fingers.)

Craig: Nine add seven that's sixteen, it's one ten and six. (But he proceeds to press a 6 and then a 1 to enter his result into the micro.)

Steven (intervening across the group): Craig, you've got sixty-one now, it's the wrong way round, one then six makes sixteen!

Kelly: Eighteen and seven is seven jumps on the number line that's twenty-five.

This may read somewhat prosaically, but there is a lot of language here for the teacher to work on. At what stages, and how carefully, for example, should we introduce the terminology for the operations of add and subtract? Note how Kelly has slipped in the ubiquitous 'and' for addition, that we met in an earlier extract in Chapter 3. 'Take away' seems universal for subtraction in the present group, as it is with many children. But again we met a group earlier who seemed baffled by 'What's the difference between . . .?' and used *counting on*, rather than *taking away*, in order to solve their problem. How should we encourage the children to make use of nought and nothing (Mark)? When should 'zero' be used and what are the differences in meaning between these three words, mathematically? Note how Steven, as reported, says *eight* numbers aloud, in doing $8 + 7$, but manages to arrive at 15 correctly. Exactly how did he use his fingers? But Craig, by comparison, gets straight to 16, from $9 + 7$, with no counting on his fingers. It looks as if he has used Steven's answer to $8 + 7$ in order to do this. Then it is Steven who intervenes to correct Craig's fine disregard for order in entering 16 on the computer. In writing down a number, you can begin from either the left or the right, but in entering it on a calculator or computer, you must work from the left. Note Steven's use of language in correcting Craig – ' . . . one then six makes sixteen'. Would place-value calculator games such as those I describe later perhaps help Craig to develop more skill and understanding of this important principle?

A top ability group – Barrie, Gavin, Nara, Javid (Two pairs)

Barrie: Seventy-three take away fifty-one – Oh that's twenty-two, I can see that right away. Three take away one is two and seven take away five is two.

Gavin: Seven tens take away five tens is two tens Barrie, not just seven take away five is two. My turn now. Seventy-one take away fifty-seven - that's units first, one take seven is six.

Barrie: No, Gavin, you can't take seven from one.

Gavin: Wait a minute, Barrie, hold your fingers up -

Barrie: Gavin, it doesn't matter, one is smaller than seven, if you've only got one you can't take seven away.

Gavin: Oh, yes, of course. I'll have to take a ten from the seven tens, I was mixed up, I know now.

(Later)

Gavin: Seventy-two plus seventy-two that's a hundred and forty-four.

Teacher: Gavin, that was quick. How did you work it out?

Gavin: Well, miss, I started with seventy, added on in tens and that came to a hundred and forty and then I added on the four units.

Barrie: I added seven tens and seven tens which is fourteen tens, that's a hundred and forty, and four is a hundred and forty-four.

Nara: Forty-one add fifty-three - fifty add forty is ninety and one and three is four, that's ninety-four.

Javid: Eleven add seventy-eight - that's eighty-nine because one ten and seven tens is eight tens and one and eight is nine.

Nara: I would have counted on from seventy-eight - it's just as easy, I think.

Javid: I think I was quicker than you, Nara.

Nara: Twenty-seven add eighty-two . . . Miss, seventy add eighty is, um . . .

Javid: No, Nara, it's two tens and eight tens then seven units and two units.

Nara: Oh yes, two tens and eight tens is ten tens, that's a hundred so it's a hundred and nine.

More interesting language - here I will leave it to the readers to test their powers as acute observers, by analysing the comments of the 'top ability group'. Although only one member of each pair is interacting with the computer directly ('My turn now . . .') the other member in each case seems to be thinking about the problem being presented and helping the partner. Probably the interaction diagram of Figure 5.4(ii) applies here, but with some degree of child interaction. It is clearly this shared thinking that corrects misunderstandings, as with Steven and Craig, or Barrie and Gavin, and not the microworld of the program. You can see the 'levels of difficulty' operating between the two groups, but also the limitations in this program. Thus 73–51 would not be regarded by teachers as at the same level of difficulty as 71–57, for obvious reasons. It is clear that they are not for Barrie and Gavin! The teacher in this case is able to use the information she has gained about the children's thinking on these problems, to guide her teaching in the next stage. *Trains* is a useful program, then, to support practice in skills, but it does no more than this. Unless the teacher gains clear and detailed feedback from the children's talk, the program may amount to little more than what Pearson disparagingly, but justifiably, calls an 'electronic child minder'.

The teacher here was the mathematics coordinator in her school. She was particularly concerned, therefore, with decisions about the most effective way to use the single microcomputer and the programs available. Each class was designated certain days for its use; it was moved around the school on a special trolley. Appropriate programs from the packs were extracted and placed onto new tapes, each designated for a particular age range. This saved class teachers having to run through whole tapes to find suitable material for their own children. The school tries to extend its program library by ordering possible additions on approval. When they arrive, all teachers attempt the program with their class and make comments on its merits, shortcomings etc. These comments are discussed, and form the basis of the decision about the usefulness of the program. In this way all staff gain a sound knowledge of the software available. The children use the computer in pairs or small groups (as in the extracts above). When the school acquires a disc drive for its BBC machine, programs will have to be transferred from tape to disc, but using them will, of course, be much less trouble.

Ergo - A Number Game

Ergo is another program from the *Micro Primer Pack*. It is aimed at upper junior and lower secondary children. For those not familiar with it, the game involves twenty-five numbers, arranged in a five-by-five grid, which form a pattern. Only two of the numbers are shown on the screen at the start of the game; the task is to discover the pattern and fill in the missing numbers. The microworld sends messages which tell the players whether their guesses are too large or too small. It will also judge whether the players have discovered the pattern, and save work by completing it for them. Problems are chosen at random from more than 1000 patterns, and offered at two levels of difficulty. The quality of this microworld may be judged from the following extracts. The first group consisted of three second-year juniors, Simon, Dale and Joanne. They played the game along with their teacher, at the 'simple' level of difficulty.

Simon: What do we have to do?

Teacher: You have to find numbers, instead of the question marks. You have two clues to start you off. Can you see ten here, and fifteen up here? You can choose by moving the arrow keys. Where shall we begin?

(The computer screen looked as in Figure 5.5 at the start. The children made no response to the teacher's opening question.)

?	?	?	?	?
?	?	15	?	?
?	?	?	?	?
?	?	?	?	?
?	?	?	?	10

FIGURE 5.5

Teacher: Let's start straight underneath the fifteen. What number shall we start with?

(Again the children made no offers.)

Teacher: Just pick a number. Come on, Dale.

Dale: Ten.

Teacher: Press ten and then Return. What is the message?

Dale: Ten is too small.

Teacher: Right, that helps us, doesn't it?

(Still there is no response.)

Teacher: If ten is too small, choose another number. Think now, ten was too small.

Joanne: Twelve.

Teacher: Good girl. You said twelve. Let's see if twelve fits the pattern.

Joanne: Yes!

Teacher: Twelve might not have been right but the message helped us not to pick a smaller number than ten, didn't it, Joanne?

Joanne: Yes.

Teacher: You try now, Simon. Pick a number that may come after fifteen and twelve. Can you see the light flashing under twelve? What do you think, Simon?

Simon: I don't know.

Teacher: I'm not sure either; have you got any ideas, Dale?

Dale: I think it's a less number than twelve. Because twelve is less than fifteen.

Teacher: Good boy, you may be right. Try it.

Dale: Eleven or ten or nine.

Simon: Try ten.

Teacher: Good, Simon.

Dale: O.K.

Simon: Ten is too large, try eight.

Dale: O.K.

Simon: Eight is too small . . . it's nine. If ten is too big and eight is too small, it's got to be nine.

Teacher: Joanne – your turn. Have a try at the next number. The one straight down from nine.

Joanne: Six.

Teacher: Did you guess, Joanne?

Joanne: No, I worked it out. Three more as you go up; three down as you go down.

Teacher: Do you all agree with Joanne? . . . Take the arrow up to the top – above the fifteen. What number goes there if we use Joanne's rule, 'three more as you go up'?

(At this point the computer screen looked as in Figure 5.6.)

?	?	?	?	?
?	?	15	?	?|
?	?	12	?	?
?	?	9	?	?
?	?	6	?	10

FIGURE 5.6

Children: Eighteen!

Teacher: Try eighteen . . . it works! Now where shall we go?

Dale: Between the six and the ten.

Joanne: It's between six and ten. Try eight . . . it is eight!

Teacher: You try now, Simon.

Simon: Go to the other side of six and make it two less. Try four . . . I was right! The next number will be two . . . I'm right!

It seems here that the teacher has to do some teaching at the start, as the children do not respond. This concerns how to use the error messages from the computer, and where to go in predicting the next number. Joanne does the first real thinking by the children and is rewarded - but by the *Ergo* microworld, not the teacher. Note how the teacher uses various *procedural interventions* to keep the thinking with the children. Study their responses and see how she draws them in more and more. She extends Joanne's response of 'Six' so that the boys learn about her method from *Joanne herself*, and then suggests testing the new rule. Then Dale takes over on where to go next, for the first time, with his 'Between the six and the ten'. In the final part, the children have taken over the mathematical thinking completely. But they might not have shared their ideas as effectively initially, unless the teacher had played her part. *Ergo*, I think, helps the children to build ideas about sequences and number patterns in a very motivating context. Its messages are clearly of a higher quality than with *Trains*. The interaction begins to approach the state of affairs I put forward as an ideal in Figure 5.4(*iii*), where the group is presented with a problem which they consider together. I think this is illustrated even more clearly in my next two accounts of groups working with *Ergo*.

(The same teacher is introducing the program to three third year juniors, who have been given instructions about what is expected.)

Philip: Eleven and one are our clues.

Gareth: Eleven is more than one, so try a number less than eleven.

Richard: Try nine . . . Good guess!

Philip: Nine is two down from eleven, so try seven . . . it's right!

Gareth: Take two away again - five . . . yes, five.

(At this stage the screen looked as in Figure 5.7(*i*))

Richard: Now we've got five and one. Five is more than one, so try a number between five and one - try three . . . it's right!

Teacher: What's the pattern?

Richard: It's two more going across and two less going down.

Teacher: Do you all think so?

?	?	?	?	?
?	?	11	?	?
?	?	9	?	?
?	?	7	?	?
?	?	5	?	1

(*i*)

?	?	?	11	?
?	?	11	9	?
?	?	9	7	?
?	?	7	5	?
?	?	5	3	1

(*ii*)

FIGURE 5.7

Children: Yes!
Teacher: Check it with some more numbers.
Gareth: Work up from three. Add two every time . . . five then seven then nine then eleven . . .
Teacher: Well done. Now stop a moment. Can you work out what this number could be – this one, to the right of the eleven?
(See Figure 5.7(*ii*))
Richard: Nine.
Teacher: How did you get nine?
Richard: Well, I looked down to the three, and three is two more than one, so eleven will be two more than nine, so it's nine.
Teacher: Check . . . Right! You are right, it's nine. Do you want to do any more?
Philip: Yes –
Gareth: Look, the nines are coming 'one up' each row.
Teacher: Does that help you?
Richard: Move the arrow to here and see if it's a nine. That's three moves down and three across . . . it is nine!
Teacher: That's a diagonal, isn't it? Try the last one in the diagonal –
Children: Yes, it's nine!
Richard: You don't need the rule now. You can just look at the pattern – seven goes in there.

The children certainly seem to be working well together on the problem presented by the program. Notice how the teacher's 'What's the pattern?' draws out a very nice conjecture from Richard, with which the other two agree. A little later she challenges them to make a 'jump prediction' rather than work step by step as they have until then. ('. . . this number, to the right of eleven?') When Richard's explanation is drawn out by one of those 'How did you get . . .?' questions, it is interesting and probably not one which we might expect. Towards the end of the extract, the teacher injects a little exposition by mentioning and then using 'diagonal'. So she is trying to adopt the new role we have outlined – trying to keep the pupils thinking, drawing out, extending and challenging where it seems possible and never pushing in when she can find a child to conjecture or explain a method! *Ergo* is again proving a very useful microworld, helping children to build and test, that valuable property of programs that I highlighted earlier on. You can see this happening again and again in these two accounts; it helps the teacher to avoid making any comments about correctness of the children's ideas. This mathematical situation is probably contributing to at least three of the areas I listed on page 76; **C** conceptual structures; **D** general strategies; and **E** personal qualities. My final account about *Ergo* shows a group of three fourth-year junior girls working together, starting from the screen of Figure 5.8.

?	?	?	?	?
?	?	22	?	?
?	?	?	?	?
?	?	?	?	?
?	?	?	?	12

FIGURE 5.8

Kimberley: We've got a twenty-two and a twelve.
Claire: Try twenty.
Shareen: Twenty is too small, try twenty-four.
Kimberley: Twenty-four is too small, try thirty.
Shareen: Thirty is too large.
Claire: Thirty is too large, twenty-four is too small. Try twenty-seven.
Kimberley: Twenty-seven is too small, try twenty-eight . . .

it's right. It's twenty-eight. Press the down arrow, Claire.

Claire: Twenty-eight is six bigger than twenty-two, so try twenty-eight plus six, try thirty-four.

Shareen: Thirty-four is right. So thirty-four plus six equals forty. Try forty . . . it's right. It's forty.

Claire: Now we've got to find a number between forty and twelve.

Kimberley: Forty take away twelve leaves twenty-eight. Half of twenty-eight is fourteen, so add fourteen to twelve. Twenty-six, try twenty-six.

Claire: Twenty-six is right.

You might like to fill in the screen yourself to follow the children's reasoning, but it is clear that they are well on their way to a joint solution. They each contribute to the mathematical thinking as they build and test their ideas about the problem presented to them by *Ergo*. They understand how to use the messages from the program, and have developed a good strategy for deciding where to make their next prediction. Last but not least, they have the social skills to cooperate in the solution. I think the nature of the interaction here certainly satisfies my ideal diagram of Figure 5.4(*iii*)!

I will conclude this section by comparing the two programs, *Ergo* and *Trains*, in terms of Figure 5.3 – ie the extent to which they enable the children to build and test ideas. We have seen that both can sustain group activity, although *Ergo* is the more effective in this respect. Looking back at Figure 5.3, I think that *Trains* does nothing to help children build their ideas (note how the first group uses a number line to support their thinking about subtraction). However it offers right/wrong feedback and the children *through their discussion* are helping one another to diagnose mistakes, correct misunderstandings and revise techniques. The teacher, by observing, listening and intervening, can also provide more expert help. By comparison, *Ergo* contributes to several aspects on the building side of my diagram. The main concept being formed is that of a number sequence. The children apply various arithmetical ideas – see particularly the group of girls. They are not only adding or subtracting like the boys in the previous account, or using ideas of greater than, smaller than – near the end, Kimberley calculates the arithmetic mean (average) of forty and twelve, without realising it! Then we have the problem-solving skills of conjecturing (Richard) and looking for patterns, and strategies of deciding where to make the next prediction in the grid. *Ergo's* feedback for the testing is more sophisticated than *Trains'* simple right/wrong. The group who most clearly build their ideas in response to this feedback are Simon, Dale and Joanne. They have 15 and 12 and Dale speculates that the number underneath is 'a less number than twelve'. After his suggestion of 11, or 10 or 9, Simon chooses 10! There has to be some quite complex logic before 9 is selected, but then Joanne 'worked it out. Three more as you go up, three down as you go down'. You can see how quickly the children get the idea of a sequence following Joanne's conjecture, and use it to fill in the bottom row of Figure 5.6 with 8 and 4. So *Ergo*, on the testing side, is helping to diagnose mistakes, correct misunderstandings and modify concepts - a fine microworld to use as a basis of group activity. But should it stand on its own? This is the theme of the next section.

Enhancing a topic

The main suggestion I wish to make is that programs need to be evaluated carefully for their appropriateness for particular age-groups, and for the extent to which they will *enhance*, rather than simply *support*, the teaching of a topic. They should fit in to existing teaching materials, or materials should be devised which achieve this, so that the program activity forms a natural part of a progression of other activities within a topic. Many commercially produced primary mathematics schemes now include programs specially devised to fit in to the topics in this way. However, there is no reason why other good programs might not be useful, provided suitable teaching materials are constructed. Some of these would be groupwork of the doing, talking and recording type I have described. To illustrate, I shall use an account given by Ball (1986) of a topic on 'Directions' for infants, undertaken by Mrs M Wagner. The software package chosen by Mrs Wagner was *Directions 1*, written by Brighouse *et al* (1983).

This package consists of three programs on left and right turns, showing that the directions of left and right are dependent on the current position. In the first program, a racing car moves round a track on the screen. As the car reaches a bend the user has to decide whether a left or a right turn is to be made.

As an introduction to this program, each child constructed a simple 'compass' (Figure 5.9). They went into the playground and were encouraged to observe and discuss their environment. On returning

Forwards
Left
right
backwards

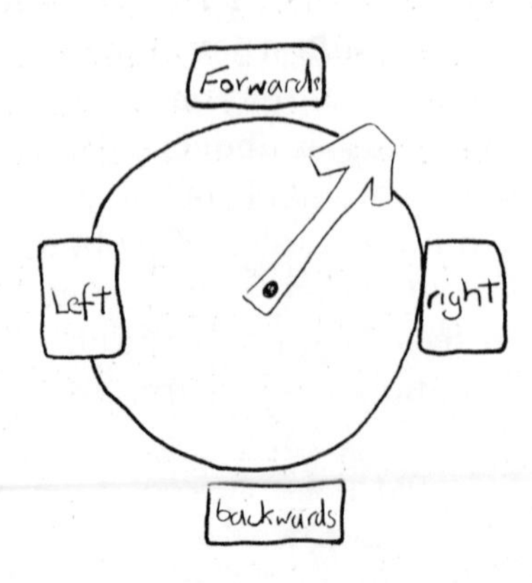

FIGURE 5.9

to the classroom, various written and drawn recordings were produced – a typical one by a six year old is shown in Figure 5.10.

A further practical activity followed, based on the doing – talking – recording icon. Children paced out routes from a given

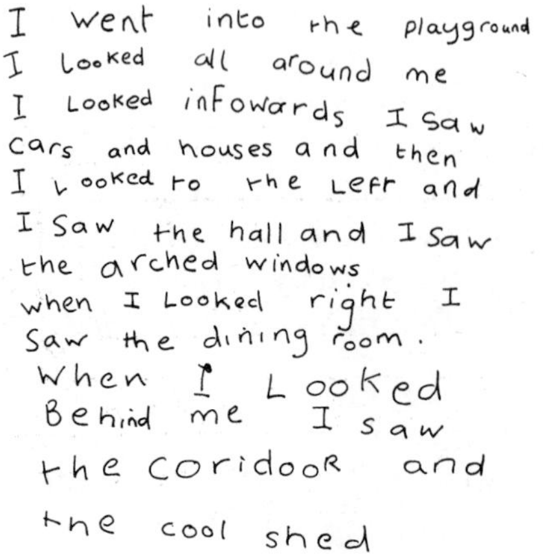

I went into the playground
I Looked all around me
I Looked infowards I saw
cars and houses and then
I Looked to the Left and
I saw the hall and I saw
the arched windows
when I Looked right I
saw the dining room.
When I Looked
Behind me I saw
the coridoor and
the cool shed

FIGURE 5.10

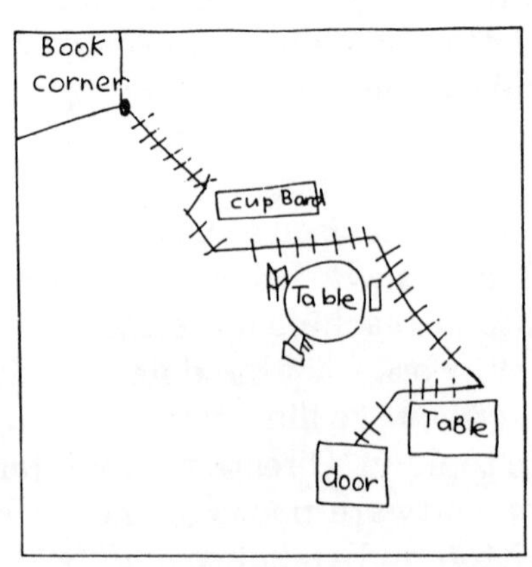

Book corner
cup Bord
Table
Table
door
I took 32 steps
if I move the Tables.
and cHairs out of the way
I take 20 steps

FIGURE 5.11

starting point to a given destination. Much discussion ensued, many alternatives were considered and recordings of the activity were made, as in Figure 5.11. The activity culminated in the writing of simple pre-Logo type programs to describe the selected route, as in Figure 5.12.

At this stage Mrs Wagner introduced the first program, called *Follow the Road*. It soon became clear that although the children's practical work showed that they could identify their left and right hand sides, whichever way they were facing, and draw their paths, they could not transfer the idea to the more abstract context of the track on the computer screen. A familiar confusion was in evidence – the children could not *imagine* which way they would be facing in driving the car round the track; left and right remained fixed across the screen, in the way that they were sitting. (Some people experience similar difficulties in map-reading.) More doing, talking and recording was clearly needed by the children, so Mrs Wagner constructed a set of work cards similar to that shown in Figure 5.13. Each child had a toy car and was encouraged to pretend that he or she was a racing driver and move the car round the track. The teacher explained that as the car reached an arrow on the track the driver had to make a decision – whether to turn right or left. The

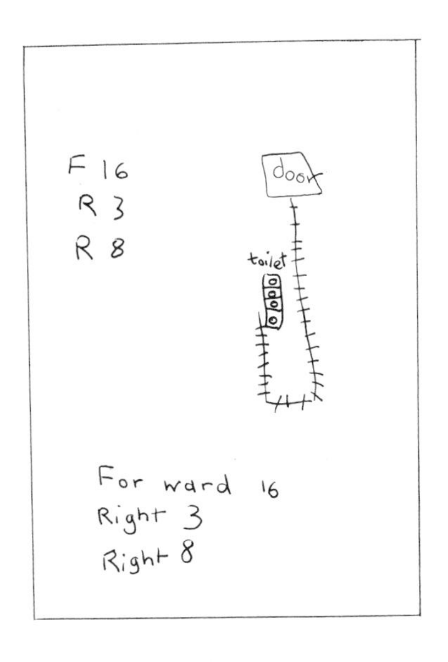

FIGURE 5.12

FIGURE 5.13

children enjoyed this activity and soon realised that it helped if the card was turned so that the 'driver' was looking along the arrow. This discovery, coupled with the physical experience of moving the car round with their fingers, seemed to help the children to understand better what was needed in the computer simulation, when this was reintroduced.

The second program in the package, *Keep on the Path*, was now introduced. It is very similar to the first, but features a man moving along a path. The figure stops at a junction and awaits the instruction to turn right or left; if the instruction is wrong, the figure crashes into a wall. Suitable sound effects and graphics in this microworld give appropriate feedback, and the children practised their newly-acquired skills with obvious pleasure and motivation. But notice it is *practising* that is reported. Would the children have gained understanding without the previous practical experience? *Steer the Boat*, the third programme in the package, permitted the children to apply their skills in a problem-solving microworld. The simulation involves steering a rowing boat into harbour, avoiding a maze of rocky obstacles. The boat leaves a 'wake' behind it as it moves forward on instruction. Possible routes into harbour have to be planned and discussed, sequences of decisions considered and modified, alternatives reviewed and compared. The children were eager and vociferous:

> *You've got to turn right but if you go too far you'll hit the rocks. Go forward a little, turn left then go forward to the top of the screen before turning right into the harbour.*

The intrinsic idea is actually quite an old one – you will find it described in an early edition of the journal '*Mathematics Teaching*', in very similar terms. But the microworld puts the idea into a very motivating form – mistakes cause the boat to hit the rocks. Thus the program quite clearly enhances the topic – once the fundamental ideas have been learnt by other means! A program based on the same ideas, but suitable for older juniors, is *Walk*, developed by Pearson (1986). Journeys have to be planned around rooms containing various objects, which have to be visited and identified. The program is accompanied by worksheets and other materials to make up a well-planned and helpful package, which could form an interesting investigation for a group.

More advanced simulations may include mathematical applications, and be supplied with support materials such as posters, pupil books, audio tapes and other materials. These programs might form the basis of a theme for a whole term's work. A good example, and one distantly related to the previous programs, is the simulation *Lifeboat*, produced by the MEP (Wales) project. It models the decision-making arising from a rescue off the coast involving the casualty, the Royal Naval Lifeboat Institution (RNLI), the Coastguard service and a Royal Air Force helicopter. The program includes a database using actual coastguard records for Swansea in 1984. Mathematical skills involved include distance, time and speed calculations, course plotting and map reading, all of which would need to be taught in related activities. This well-produced microworld allows children to participate in an activity which would be quite impractical for them in real life. The potential for other areas of the curriculum is obvious.

The micro as a tool

So far I have described examples in which the first two uses offered by *Mathematics 5–16* (as a teaching aid and as a learning resource) have appeared. The situation when children use the micro as a tool is rather different; in this context, the diagrams I introduced in Figure 5.4 need modification. Instead of thinking of the microworld as offering a problem which the group consider together, the problem under consideration is external to the program, which functions as a means of solution. The children are likely to be applying ideas, rather than building them (in Skemp's sense). Nor

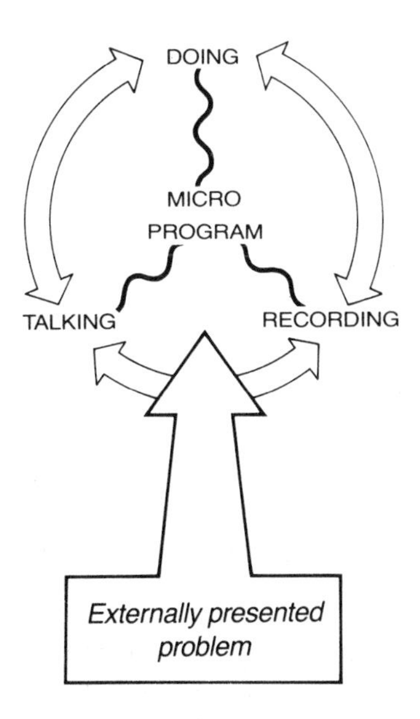

FIGURE 5.14

will a program used as a tool necessarily offer a means of testing ideas – perhaps only a way of developing them. For example, a three-dimensional graphics program might enable me to design a kitchen – where to place fittings and so on – by comparing different arrangements quickly. But judgements about the best arrangement would be made by me, not the program. On the other hand, a simulation program might enable an engineer to design a car shape with minimum drag, by testing different shapes. Again, many of the judgements, which usually involve compromises, will be made by the designer. The use of the micro as a tool, therefore, involves a significantly different situation from those so far described. Figure 5.14, a modified version of Figure 5.2, is intended to suggest this.

Information handling and word processing are the two areas where it seems most likely that children would be using the computer as a tool in the sense used here. Clearly there are some mathematical aspects of data collection and entry, but the use of programs as a *problem-solving tool* at the primary stage, in the manner of scientists or designers, probably remains rather small at present. Pearson (*op cit*) describes two areas of opportunity. The introduction of Logo creates an 'authentic problem-solving environment' in which this powerful language might be used by children to develop their own programs in order to solve problems. He mentions the use of Logo's list-processing capabilities to design adventure programs as an example. The second area is that of Craft, Design and Technology, where there are arguments for the early introduction of aspects of robotics and computer control. These are likely to involve the construction of control programs, where once again Logo seems the best available language. All these developments have been well described elsewhere, and in any case await the outcome of research and resource decisions before their introduction becomes general. My own example is much more modest, and more typical of the present stage, I think.

The teacher concerned used an idea put forward by Buck (1986) to develop an investigation into digital root sequences with a group of six fourth-year juniors of above average ability. (The 'digital root' of a particular number such as 237 is found by adding its digits together repeatedly until a single digit is reached, eg in this case 237 ⇒ 12 ⇒ 3, or 145 ⇒ 10 ⇒ 1.) To start off, the group was asked to say the three times table while one member wrote the responses. Next the teacher explained the meaning of digital root of a number, as above, and the group recorded in a further column the roots of their three times table –

3	3
6	6
9	9
12	3
15	6
18	9
21	3

and so on. Now the teacher explained how to use this to draw a pattern on 'spotty paper', an idea developed by Buck. The result is shown in Figure 5.15, using the rule that a clockwise direction of turning is to be maintained. At this point the group was asked to investigate more patterns of digital roots using different number sequences. The task of drawing them was given to each member in turn.

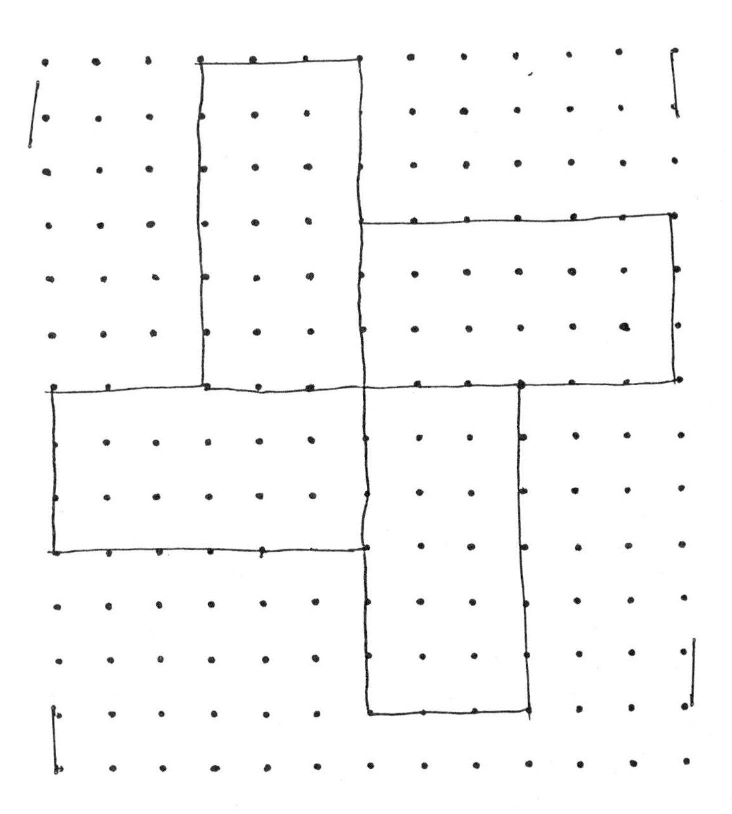

FIGURE 5.15

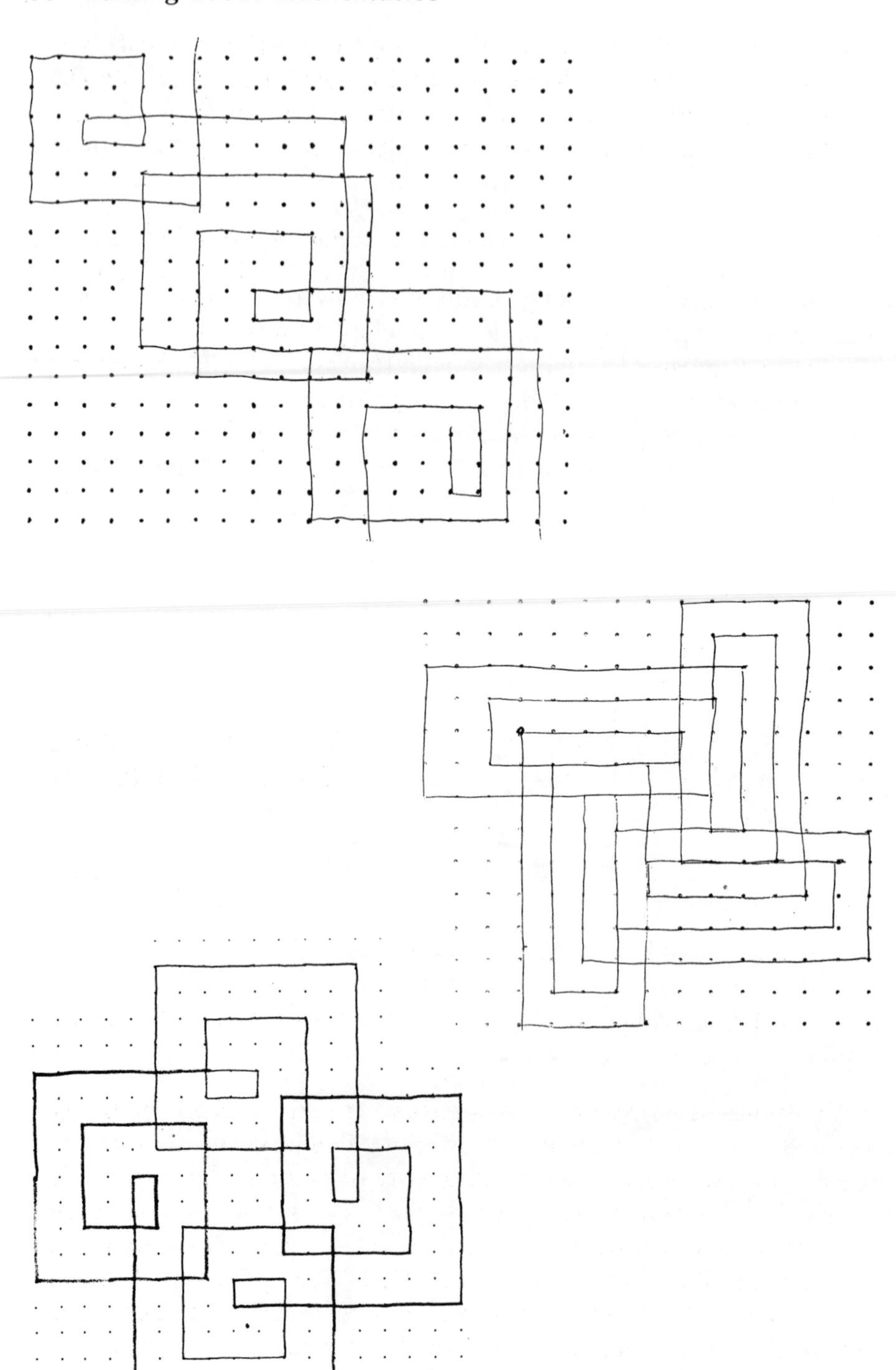

FIGURE 5.16

The situation generated much discussion throughout, most of it centring on predictions about emerging patterns, with all the children actively involved. The mathematical outcomes were striking - several of the drawings are shown in reduced form in Figure 5.16.

The group found that the root sequences repeated after 9 numbers, except for the 3, 6, and 9 times tables. Their overall results were presented in the form of a table:

Times table	Digital root sequence								
1	1	2	3	4	5	6	7	8	9
2	2	4	6	8	1	3	5	7	9
4	4	8	3	7	2	6	1	5	9
5	5	1	6	2	7	3	8	4	9
7	7	5	3	1	8	6	4	2	9
8	8	7	6	5	4	3	2	1	9

The 3, 6 and 9 times tables gave sequences $3 \Rightarrow 6 \Rightarrow 9$, $6 \Rightarrow 3 \Rightarrow 9$ and just 9, respectively. Several questions were formulated for further investigation:

1 Why do the patterns for 1, 3, 4, 5, 6 and 9 seem 'complete', while those for 2, 7 and 8 'go off the paper'?
2 Why was 9 the last digit in every sequence?
3 Why were 3, 6, and 9 different from the other patterns?
4 What if triangular spotty paper were used instead of square paper?

At this point a short program in Basic given by Buck helps by acting as a powerful drawing instrument. The program accepts the digital root sequence as input and then draws the diagram. It is designed to rotate the main pattern four times, to produce a more interesting final result. Another continuation is to get the children to construct their own 'tool' by writing a Logo program. Logo is particularly well-suited to this problem, and helps to show that some of the 'off the paper' sequences do close in time - Figure 5.17 shows the completed two times table pattern.

It is reasonable to argue that in this case the children are using the micro as a tool, and rather differently from the other groups I have reported. It seems likely that we would find this happening with the use of investigations in the primary school. Once again these would

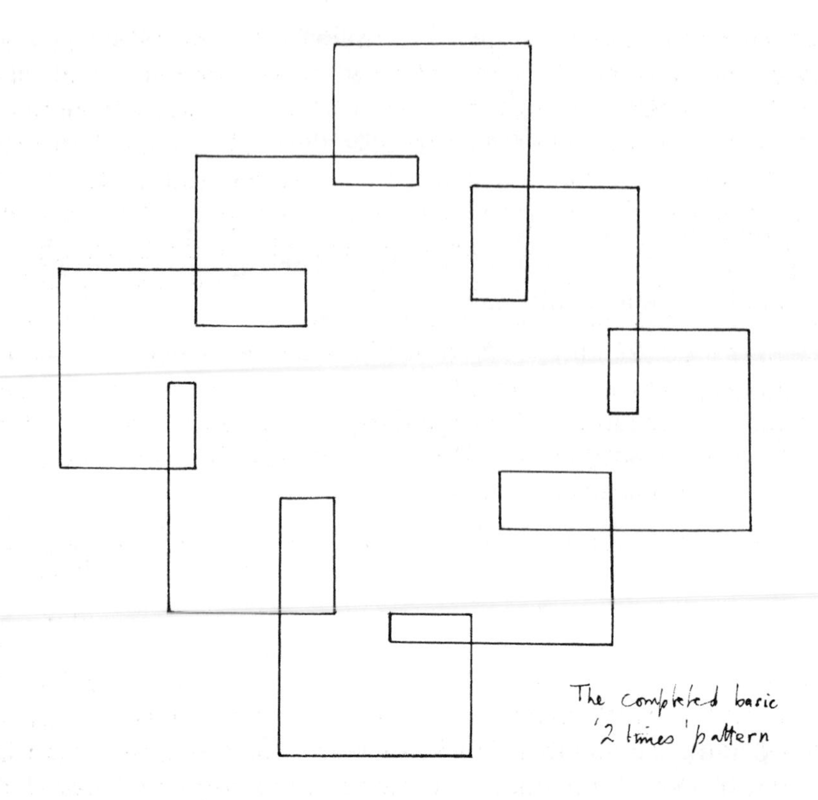

FIGURE 5.17

provide an opportunity for small groups to work together on the problem. What kinds of programs might be useful to primary children in this 'tool-using sense', and who would write them? Programs which operate with whole numbers suggest themselves, for use in the number investigations that are likely to form a major part of this aspect of the mathematics curriculum. Perhaps these might link with the calculator situations that form the theme of the next section. Straker's *Picfile* (1984) is a suite of programs which can be used to construct graphical displays of data. But clearly, the field seems ripe for continued development.

Calculator conversations

Mathematics 5–16 discusses the use of calculators under Objective 6, and two key sentences stand out from the discussion. The first, referring to a statement that 'allowing pupils to use a calculator is not sufficient', goes on to say that what is needed is a policy which 'encourages pupils of all ages and abilities to use calculators in appropriate situations and provides clear guidance on the procedures needed to obtain maximum benefit from their use'. Later it is said that '*only very basic and simple calculations now need be done on paper*; some standard written methods of computation, such as long division, which many pupils find difficult and few really understand, should no longer be generally taught'. I think we can see the calculator in the primary classroom playing exactly the same three roles as the micro - as a teaching aid, as a learning resource and as a tool. Primary children, indeed all children, are very much more likely to use the calculator as a tool than they are the micro. A particularly important use of this kind would be in investigations. But it could also be very useful to teachers in the other two ways.

As a learning resource, the calculator could form part of a doing, talking and recording situation such as I have described in Chapter 3. Calculator games are the obvious line of approach, and there are many useful sources of these. A key reference, both on this and on the other aspects, is *Calculators in the Primary School*, published jointly by the Mathematical Association and the Association of Teachers of Mathematics (1986). Section 4 of this publication is devoted to calculator games (and many others are scattered through the articles). Another useful booklet is *Calculators*, published by Leapfrogs. I shall describe a group of fourth-year juniors playing a game from it. A 4 × 4 grid of numbers is displayed to the group on a large sheet of card, as in Figure 5.18.

174	138	382	230
206	170	414	262
275	239	483	331
117	81	325	173

FIGURE 5.18

Teacher: Draw a four by four grid like this . . . look at the pattern of numbers - choose four numbers from the pattern and mark them in on your grid like this to show which you've chosen . . . choose the four numbers to get as close to one thousand as you can when they're added, but you can't use your calculator or write on the paper - you have to work in your head.

The children draw their grids and shade in their choices, with various comments – *I've got my four!* – *There's my first two ...* (The five drawings are shown in Figure 5.19.)

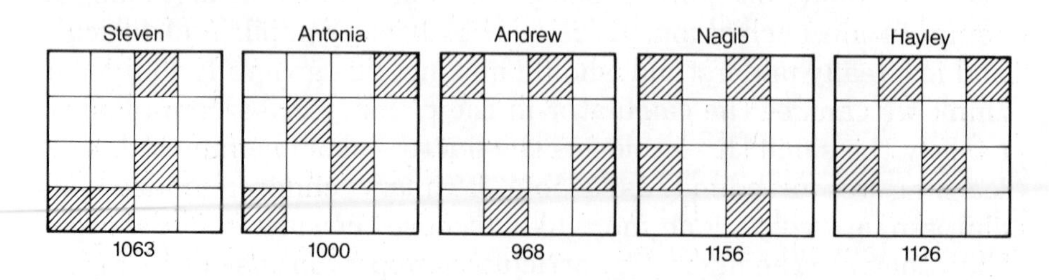

FIGURE 5.19

Teacher:	You've got to commit yourself – when you've done, use your calculator to add up to the four numbers you've choosen and write down the total. The winner is the one nearest to a thousand.
Andrew:	Ooh, she's dead on!
Teacher:	What did you get, Andrew?
Andrew:	Nine hundred and sixty-eight.
Teacher:	What about you Steven?
Steven:	One thousand and sixty-three.
Teacher:	What did you get, Hayley?
Hayley:	One thousand one hundred and twenty-six.
Teacher:	How did you get on, Nagib?
Nagib:	One thousand one hundred and fifty-six.
Teacher:	Right, now let's just think about this – who's the closest?
Children:	Antonia!
Teacher:	Right, how did you get on, Steven?
Steven:	One O six three ...
Andrew:	I'm second closest – nine six eight –
Teacher:	You're the second closest – er, thirty-two away then. Hayley, how did you get on?
Hayley:	I'm a hundred and twenty-six out!
Teacher:	So Antonia wins this round, you're second, Andrew.
Andrew:	We can use that way now!
Teacher:	All right then, we rule out that pattern, don't we! How did you do it, Antonia?
Antonia:	Just added them up, that's all.
Teacher:	What do you mean, you sort of looked at the numbers and –

Antonia: Well, just on the end, that's seven and three equals ten, that sort of adds up to six hundred, then I looked for four hundred and I just got those two.

You can follow Antonia's reasoning in the two previous figures. She proceeded to astonish the group by winning the next round with another 1000! She chose 230 and 170 again, and then found another two to give a total of 600 – I wonder if you can spot them? This game produces some talk, some estimation in a rather cunning way, and of course calculator practice in addition. But there are many other games which will prove equally effective, for example the game for four players called 'Count me out', described in *PrIME Newsletter* No 1 (1986). This teaches place value, and uses the calculator as a learning resource rather than a tool, as in the previous game. Each player has a calculator and keys in a 3-digit number, without letting the others see it. Amy asks Josh, seated on her left, 'Please give me your fours' (or any other digit she chooses from 0 to 9). Suppose Josh has entered 143 on his machine, he must say the value (forty or four tens) and then subtract it from his calculator total. Amy adds the 40 she has 'won' to her calculator total and takes another turn as she has won some points. If Josh has no fours, it is his turn to ask Javeed. Players take turns like this until one player has 'won' all the points and the others have zero on their calculators. Accuracy can be checked by all the players making a note of their original numbers. At the end of the game the winner's total should of course equal the sum of these numbers.

The calculator as a 'Black Box'

Because the calculator is a machine with built-in laws, it can be used as the basis of a 'black box' mathematical situation, in which the children are invited to explore the 'black box' in order to find out how it works. This approach proved very effective many years ago in introducing children to the old-fashioned mechanical calculating machines. It can be successful in the same way for introducing older juniors to the more powerful facilities on an advanced calculator, such as might be used in the secondary school. In particular, they can explore the use of the memory and other functions. Although they will not learn about the more advanced functions until later, they can appreciate how the calculator acts as a 'function machine' and gain an *understanding* of at least the square/square root facility. I shall conclude this chapter with a brief account of infants meeting the function machine concept, and then the same group of fourth-year juniors finding out about square roots.

The infants were 4–5 year olds, their teacher was working on the theme of 'colour', identifying, for example, 'sad' and 'happy' colours, and using the micro adventure game *Granny's Garden*.

Teacher: I want you to think about all the different types of machines that you have seen in school, in your home or outside. Can you tell me about them?

(The following comments are typical of the discussion.)

Christopher: A 'grabbing machine' – you put money in and you press the button and if you let go of the button it doesn't grab anything but if you do keep your hand on it, it grabs something.

Teacher: What about you, Ryan, have you seen any machines?

Ryan: You put money in and you press the button and a ticket comes out and then you go over to your car and stick it on your car.

Steven: When I go over to the sports centre, I go to the hair dryer to dry my hair and after the pool daddy puts money in and the air comes out.

Teacher: We've been talking about machines that we've seen outside, this is a machine that we've been talking into. What is happening to this machine?

Peter: You talk and then out comes your voice.

Notice that the kinds of machines that concern these young children are not the sort that adults, or older children, might mention and the teacher uses their ideas as far as possible. They continue with a game involving a cardboard 'function machine'. *Logiblocks* are placed in the machine's input tray and appear to have changed colour when 'they' emerge in its output tray. Similar games can be devised in which function machines change other attributes such as shape, size or thickness. (A rather different idea is the use of 'gates' which allow *Logiblocks* with a particular attribute to 'pass'.) These materials, available from commercial suppliers, were originated by Dienes, who has written about them in various sources, eg Dienes (1971). (See also the account in Chapter 6.)

From such situations quite young children can build up the idea of a function machine which accepts an 'input', operates on it in some way and then 'outputs' a result. This idea is important later on, as is shown in my final account with a group of fourth year juniors. The teacher gave them each an advanced scientific calculator, the Casio *fx-7*. His initial instruction generated ten minutes or so of interested discussion and exploration, from which typical comments are reported.

Teacher: I want you to find out as much as you can about how they work – about the buttons and so on. Talk to one another if you want to.

There are buttons to say whether you divide, add, etcetera.

You can only have an eight-digit number.

Has it got a memory on this? Yes –

Yes, you can memorise.

All calculators you can memorise on.

Most of them, anyhow.

There's no percentages.

Yes there is. (The percentage facility on the Casio is in an obscure corner.)

What is MODE?

You've got decimals on it – a point.

E – X – P . . . if you put thirty, or thirty-two – then when you press it you get three point –

When you take away it goes higher!

Why do we keep on getting that number – when we press EXP? (The number is π of course.) *It's weird.*

Teacher: You'd think if that number was built into the machine it must be very important! It's one of the most important in mathematics, actually.

(The teacher uses comments by Andrew to focus on a particular function.)

Teacher: Now here's something you can try with these white buttons – start from scratch again. Now put nine in and then press this button.

Yes – Yes – three.

Teacher: Try it with twenty-five.

It should come to twenty-five – five rather.

Teacher: It should come to five? . . . What happens if you put in a hundred then?

Ten!

I know what this is doing – it's a square number – square numbers.

Teacher: O.K. – what happens if you put in a number like fifty-seven then?

Let's try it.

There'll be a remainder of something, probably.

Ooh! Hey! (With great surprise.)

Teacher: What do you get?

Seven point five four nine eight three four four.

Teacher: All right – now write down that number . . . now, you've still got that number in the calculator haven't you? Press times . . . and then enter what you've just written down . . . and then the equals sign . . .

It's the same number!

What number did we put in first?

Fifty-seven!

Teacher: Try that with some different numbers and notice what's happening – I'll ask somebody to explain it later.

You just get back to the number you started with!

(A further period of exploration follows, in which the teacher introduces the phrase 'square root' and encourages the children to use 64 as a starter.)

Teacher: Now, I think our first number was fifty-seven, wasn't it, Hayley? And you got seven point five four nine eight three four four. What do you think that number is?

What'll go into it?

That's a square number that'll go into it.

(There is more investigation using the children's numbers, which shows that they can 'always find a number which works like that'. The teacher introduces the idea of estimating the size of the number that will appear.)

Teacher: Now I'm going to set you a problem – I'm going to suppose that those are a button that we press, some number has got to go in there that we enter. And the same one there – and when I've done this I've got to get – would someone name a number? (Andrew offers 190.) I want you to find what number that has to be.

(As he speaks the teacher points to the 'buttons' he has drawn as in the diagram below.)

$$\square \times \square = 190$$

Thirteen point seven eight four Oh four nine.

Teacher: And you're going to have to prove it to me in a minute. That was quick, wasn't it? . . . It works? . . . Everyone tried it? . . . Now what did you actually do, Antonia, to solve my problem?

I entered a hundred and ninety in then used that button. Then I wrote down the number that I had, then to test it I timsed it by itself and it worked out.

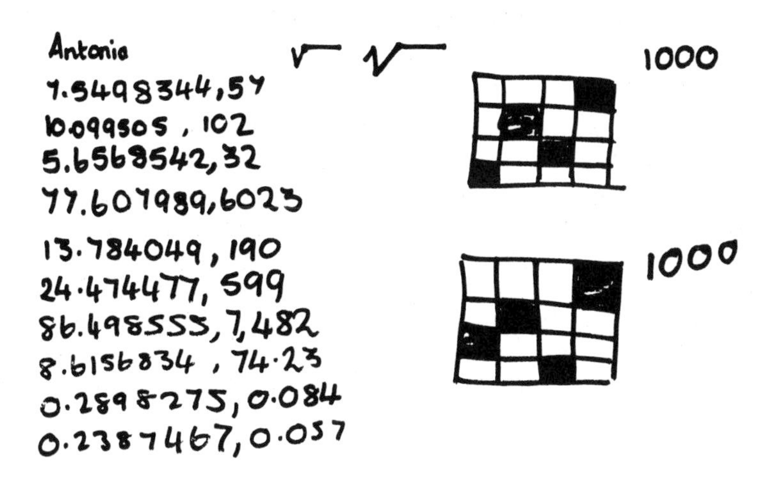

FIGURE 5.20

The investigation concluded with further work on numbers like 0.084 and 0.057, as well as the introduction of the sign for square roots (See Figure 5.20). Those who recall the days when children struggled with square root tables and all their difficulties will be in no doubt that the calculator enhances the teaching of this topic. It places more advanced topics like Pythagoras within the reach of the average child, and perhaps, with approaches as in the above account, improves their understanding of the concept of a square root. Notice how the children were able to relate their previous work on square numbers to what the mysterious white function button was doing through the use of discussion. Their early comments also suggest a considerable knowledge of calculators, although no formal teaching of their use had been done in the school at that time. The button idea used later on in the extract suggests that calculators could help with the teaching of algebra, through numbers hidden in the memory. I will conclude with a calculator game of this type for two children, which might be tried.

A calculator memory game

The children need a single calculator with a memory, and the associated buttons for memory recall and memory add (MR and M +). Suppose the children are Ranjit and Tom. Ranjit enters a number less than 50, adds it to the memory and clears the display, concealing the process from Tom. Tom gets the calculator, enters a number larger than fifty and passes it back to Ranjit. Next Ranjit subtracts his memory number from the display and returns the calculator again. Tom has the problem of finding the number that Ranjit has hidden in the memory, knowing his own number, of course.

The game teaches the use of the calculator's memory facilities and gives children practice in subtraction, with an equation-solving element. If the rules are varied to allow any numbers as input, negative values can appear. In spite of what Barrie said to Gavin in one of the *Trains* accounts ('If you've only got one you can't take seven away') the calculator seems to allow it. What does it mean?

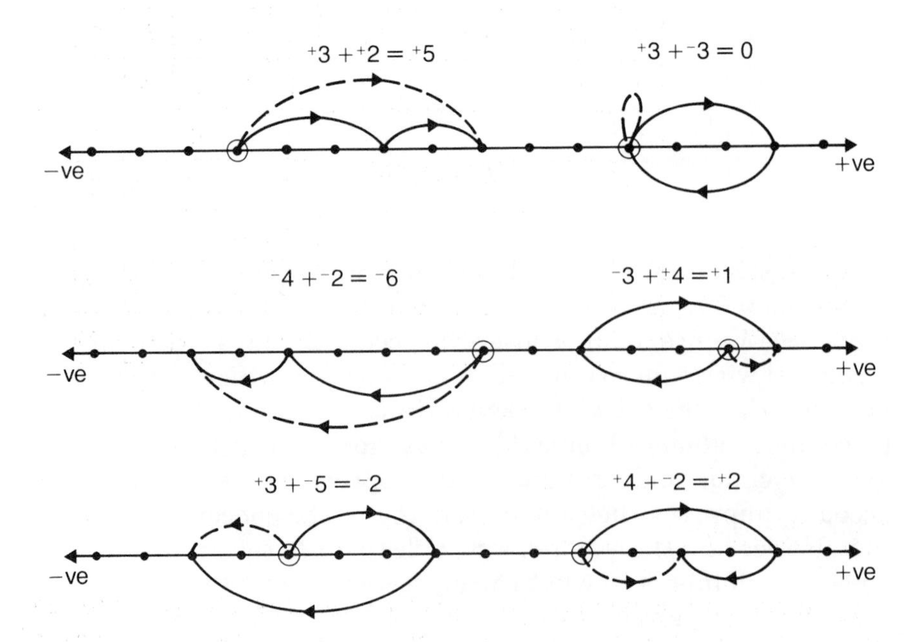

FIGURE 5.21

Presenting a problem like this is a possible introduction to the integers (directed numbers), using the calculator once again as a 'black box situation'. These new kinds of number that emerge from sums like 5-6, 5-7, 5-8, etc, can be created by using the 'sign button', denoted by ' + / – ' on the Casio and many other calculators. What happens when we add them and subtract them? When we multiply them? The addition operation can be made meaningful by using the number line in a fresh way, as in Figure 5.21. Multiplication is more difficult, but there are several situations available to back up the calculator experience, for example the well-known diagram of Figure 5.22. Integer games using the calculator could replace, or back up, the familiar pages of exercises on the 'rules of signs' when the integers are introduced.

X	−5	−4	−3	−2	−1	0	1	2	3	4	5
5	−25	−20	−15	−10	−5	0	5	10	15	20	25
4	−20	−16	−12	−8	−4	0	4	8	12	16	20
3	−15	−12	−9	−6	−3	0	3	6	9	12	15
2	−10	−8	−6	−4	−2	0	2	4	6	8	10
1	−5	−4	−3	−2	−1	0	1	2	3	4	5
0	0	0	0	0	0	0	0	0	0	0	0
−1		4	3	2	1	0	−1	−2	−3	−4	−5
−2				4	2	0	−2	−4	−6	−8	−10
−3					3	0	−3	−6	−9	−12	−15
−4						0	−4	−8	−12	−16	−20

The patterns suggest how to complete the diagram

FIGURE 5.22

Summary

The ideas and accounts put forward in this chapter are not intended, needless to say, as a definitive guide to the use of microcomputers and calculators in the primary school. I have tried to establish that they can form the basis of very effective mathematical situations, which will generate doing, talking and recording and so enable groups of children to build and test their ideas together. I have distinguished between a number of different

PART THREE
THINKING MORE DEEPLY

6 Planning effective mathematical situations

A wealth of ideas is available nowadays to help teachers to generate mathematical thinking in their children. Unfortunately, for the most part this work, as we saw in Chapter 1, is apparently either on an individual basis or takes place (rather more rarely than in the past) in a class lesson of traditional form - that is, of exposition and question and answer. The possible benefits of children's talk are thus largely lost. Probably the majority of my readers will be using one of the excellent modern commercially produced mathematics schemes as the main basis of their classroom activities. The ideas put forward in Part Three are not intended as a comprehensive guide to the setting-up of a primary mathematics scheme. Rather, they offer suggestions for the introduction of talk and discussion into the activities, and outline the factors to be looked for in effective mathematical situations which generate and support such discussion. What we suggest, in brief, is that teachers look for possible starting points within their own scheme - very often these are indicated in the materials themselves. Records can be kept of the trials of these starting points, somewhat on the lines suggested in Chapter 4. These ideas can then be incorporated into the mathematics scheme, so that other staff can draw upon them - a possible format for this will be given later. In addition, supplementary ideas can be found in various sources and these can be built up into a school 'resource bank' in a similar manner.

The microcomputer is a particularly important resource, as I described in Chapter 5. Again, sharing good ideas will be essential; many potentially useful programs need to be linked with points in the mathematics scheme where they can best be employed, and where other material fits around them. There is also the interesting possibility of having programs specially written for a mathematical situation devised by teachers. These programs might be of the microworld type, used as a learning resource, or of the 'tool-using' kind that I distinguished earlier on. But my main interest lies in those mathematical situations, of whatever kind, that form a basis of talk and discussion. I shall concentrate first on their aims and purposes, using once again the framework given by *Mathematics 5-16*.

Aims and purposes

Our discussions should be *purposeful*, with the various activities being linked with objectives. However, admitting flexibility through the use of children's own ideas raises some important issues. Mercer (1985), in an important reference to which I shall return in Chapter 9, draws attention to clear discrepancies between what the teacher intended, in various activities, and what the children afterwards thought they had learnt. He raises the questions:

1 Is it important that young children understand the purpose of what they are doing in class?
2 Is it important or useful to discuss such things with children?

Hislam, in a case study in the same reference, describes carrying out a project on woodlice with infants. He reports that

> there was a real discrepancy between my caption for the display 'What we have found out about woodlice' and what the children actually perceived the results of their work to have been . . . (they) did not actually follow the line of enquiry I had planned and in the event only came to complete the task by a rather circuitous route.

Hislam felt that, as a teacher, he had to have some overview of the curriculum and the relevance of each piece of work within the total framework. But it did not seem necessary to explain why they were doing the work in this case, as the children's own interest was reason enough. However, Hislam did discuss such aspects with older children, who he thought were interested in the reasoning behind what they did.

I think it reasonable that there should be such apparent discrepancies between the teacher's purposes and the children's perceived purposes. These discrepancies will depend on the age and sophistication of the children. But we have to go beyond this and admit the possibility of real discrepancies between what was intended and what in fact happens. If we are not willing to allow this, then discussion, in the sense I defined in Part One, is not likely to occur. You can see more than one hint of this possibility in the accounts I have given in Part Two. If our mathematical situations are effective, and *we are willing to let them work for the children and for ourselves*, then understanding will be achieved, albeit by varied routes. I shall try to define a range of possibilities, from very open to very closed situations, by adapting the framework of *Mathematics 5–16*. This begins by referring to facts and skills, then conceptual structures, and finally general strategies and personal qualities. These five categories of aims are linked in Figure 6.1 with a range of approaches, that with which they are predominantly associated being shown by a heavier line.

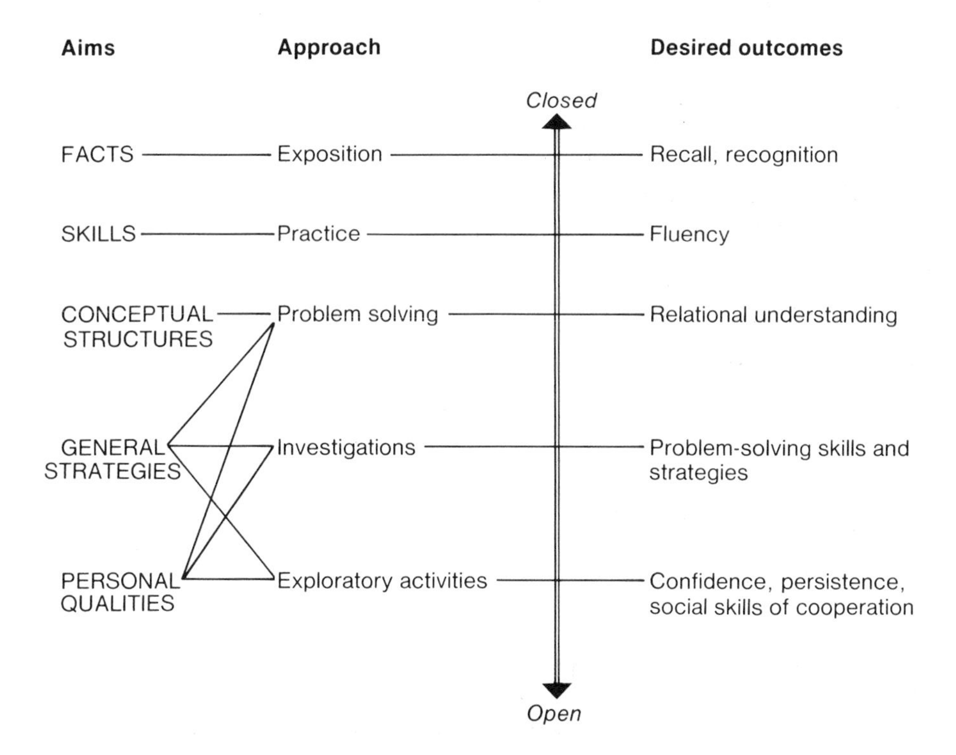

FIGURE 6.1

Talk and discussion will play a part in all of the approaches I have listed – even under 'practice', for instance, diagnostic discussion is likely to occur. But for all of the first three categories of aims, I think it can be argued that teachers do want specific outcomes, although the routes to them will be varied, because of the use of problem solving and discussion which employ children's ideas.

Control over these outcomes will be exercised partly by careful planning of the mathematical situations and partly by the use of more closed questions by the teacher in generating the problem solving. However, as we move towards the last of these categories, linked with investigative and exploratory approaches, the discussion is likely to become more open – the teacher is not directly interested in some specific skill or concept as an outcome, but in the development of strategies and personal qualities. The activities are thus wide-ranging in their use of children's ideas, feature open questions by the teacher and fewer interventions of a mathematical type. What I have termed 'exploratory activities' may not in fact be in mathematics at all, but in some other part of the curriculum. The experience gained will be used at some future point to help in learning mathematics. I shall illustrate these ideas by examples a little later.

What I aim to convey is that there is an 'open–closed' dimension to the use of talk and discussion. 'Skill' and 'concept' activities will have fairly specific outcomes, in the form of 'content objectives', although, as I said before, these might be achieved by varied routes through the use of children's ideas in problem-solving discussion and over a period of time. 'Investigative' and 'exploratory' activities will have very varied mathematical outcomes, and make freer use of children's ideas. The objectives involved will be of the kind termed 'process'. I do not believe that talk, and particularly discussion, is tied only to the open end of this dimension, although some may differ over this. The difference lies in the fact that in the first case the goal of the activity (and the discussion) is defined, however implicitly, by the teacher. In the second case the goals are largely defined by the children. This is suggested in Figure 6.2.

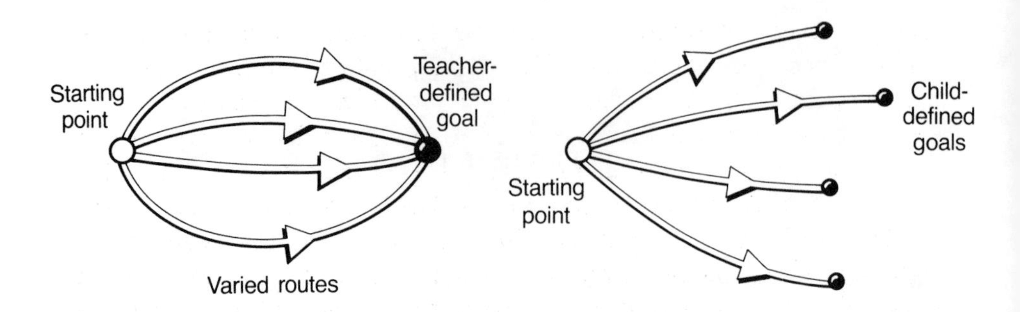

FIGURE 6.2

More about DO-TALK-RECORD and BUILDING and TESTING

I now want to develop the ideas which I introduced in Chapter 2 in rather more detail, since they have an important bearing on the way in which we choose and plan our mathematical situations. First I remind you of Skemp's distinction between instrumental and relational understanding which appeared back in Chapter 1. The instrumental approach to teaching mathematics emphasises the manipulation of symbols in the form of 'This is how you do it' – the learner is simply asked to accept rules and definitions from the teacher, and to imitate as closely as possible how they are applied. There is no discussion of why or how the particular rules came into being, or of why they happen to work, and little or no attempt to 'negotiate meaning' for them. Skemp's relational understanding, on the other hand, depends entirely on approaches which develop meanings around the rules and the symbols which express them. The learner is never asked to accept, but instead to try to develop her or his own meaning. Because mathematics is ultimately expressed as a symbol system, this process of negotiating meaning is particularly difficult. This is why doing and talking are so important as components in the process. James and Mason (1982) comment

> The struggle to capture an insight which is as yet pre-articulate is often overlooked in a rush to lead students into formal symbols, resulting in an impoverished if not empty background experience, and producing frustration, anxiety and math-phobia.

The psychologist Bruner (1973) has given a detailed account of the development of symbols. He distinguishes three modes of representation by children – the enactive, in which past events are represented by motor responses, the iconic, which uses images in a way which closely models past events, and the symbolic, in which conventional sounds or signs are used with associated meanings. Clearly these modes are increasingly remote from the events they represent, somewhat like the difference between the actual experience of some event, a picture of it and a headline about it in a newspaper. An example of this transition can be seen in the cube game I described in Chapter 3. In the game itself the children were actually making the shapes (enactive) but the later activity of drawing their shapes on isometric paper is a form of iconic recording - very closely modelling the originals. I myself took a photograph and also drew plan views of their shapes on squared paper, shading in the squares where the two blue cubes had been placed on the five yellow squares representing the base. The lower ability group in the account *Trains* in Chapter 5 are actually making use of an iconic representation - the number line - to perform their calculations. The infants in the 'Unifix steps' account might make an iconic recording of the outcomes of their enactive stage, which produced their sets of rods, by drawing them, as I did in Figure 3.4. The names attached to them by the children when they talked involve the use of symbols - the word 'three' has only a conventional link with the corresponding rod. In the account, Leanne shows that she has established these links between the sequence of spoken symbols and the set of six objects she has made (cardinal meaning). She negotiates with the other children to establish a further (ordinal) meaning by making the steps.

Bruner's ideas have been made use of by psychologists studying the learning of mathematics. The general thrust of their work is that children need more opportunities to develop meanings linked with mathematical symbols, and this might be done by planning situations involving Bruner's enactive and iconic modes, rather than heading directly for formalisms. James and Mason (*op cit,*) have developed this approach in some detail, and it is embodied in the Open University course *Developing Mathematical Thinking* (EM 235). From them I have adapted the icon of Figure 1.2, based on DOING, TALKING and RECORDING. We saw when I discussed the ideas of Vygotsky and Halliday in Chapter 1 that learning to speak not only enables the children to get things done, but, by affording them symbols, significantly changes their powers of introspection. Skemp uses the notion of 'reflective intelligence' and says that it is by means of symbols that we achieve voluntary control over our thoughts. Speech precedes writing and reading, which develop along with mathematics at the primary stage, and it seems almost obvious than we should make far greater use of this well-developed ability rather than relying so much on the others. Effective mathematical situations, if we accept these arguments, need to include work with apparatus of various kinds, discussion between children themselves and talk with the teacher. The children are encouraged to articulate their thoughts to one another, and to developed their own informal recordings. From these informal recordings, by a process James calls 'successive shorthanding', the usual symbolic forms are developed. If this approach is adopted, it follows that the emphasis on exact following of steps in a rule, using symbols prescribed by the teacher, is abandoned. We expect the children's forms of recording to appear in their written work instead, and probably a greater use of drawings and diagrams. It does not follow that clarity and neatness are abandoned, of course!

James and Mason, in describing their theory, say that the classroom activities of doing, talking and recording facilitate 'corresponding shifts in psychological states'. These shifts are from enactive to iconic, in which the children manipulate instances and get a sense of some generalisation, then from iconic to symbolic, in which articulations of the generalisation are crystallised into a recording (of their own) which captures the idea for them. Their final shift is from symbolic to enactive – that is, the idea now becomes a kind of 'mathematical object', to which the corresponding symbols refer. Children with a well-established meaning can manipulate this new 'object' as if it were in the real world. The older juniors in my extracts are often manipulating numerals in this way – for example the group in my account of 'What Does "And" Mean?'. My icon of Figure 1.2 focuses on the classroom activities to be observed, rather than the psychological theory, since I am mainly concerned with bringing about the necessary conditions for talk to take place.

Skemp (1982) writes in a similar manner about the importance of talk. He points out that 'The connections between thought and spoken words are initially much stronger than those between thoughts and written words or symbols. Spoken words are also much quicker and easier to produce.' Skemp, like James and Mason, believes that the use of informal, transitional notations would help to produce better understanding. By allowing children to express their thoughts in their own way to begin with, the recordings used are already well attached to their associated concepts. These recordings may be lengthy and unclear and may also differ from one child to another. The teacher in the extract 'What's the Difference?' reported this when her group tried to record their attempts to count on using Lee's method. Lee's own recording (Figure 3.6) is unlikely to be clear to non-participants in the discussion. Skemp suggests the use of discussion about the advantages and disadvantages of these informal recordings, so that when the teacher introduces conventional forms the children have a clearer idea of the reasons for the choices that have been made in adopting them. In this connection you might like to read again the section in Chapter 1 on 'Talk as a means of improving language skill'. The extract entitled '1, 2 and 3' in Chapter 4 shows such a process, I think – the children liked Ryan's idea and their recordings all revealed his influence, although I have only reproduced some of them.

I now want to describe in more detail Skemp's building and testing framework that I linked with the observed activities of doing, talking and recording in Chapter 2. 'Schemas' is the term commonly adopted to describe the mental structures which give meaning to symbols, enabling us to think and act with them. We are able to operate on our own schemas by reflection, Skemp argues, so changing the ways in which we respond to demands on us or choose to act ourselves. This process of construction or modification of our schemas has the twin aspects of building and testing. Moreover, there are three modes in which such building and testing can occur, giving altogether six forms, each of which Skemp terms a *functioning*. His overall picture is shown in Figure 6.3.

FIGURE 6.3 Functionings in the construction of schemas

Building	Mode	Testing
From experience (acting in the real world)	1	*By experiment* (testing our expectations of events)
From communication (using the schemas of others)	2	*By discussion* (comparing ideas, experience with others)
From creativity (forming fresh concepts by extrapolation,intuition)	3	*By internal testing* (comparing one's own knowledge, beliefs, seeking self-consistency)

In Skemp's mode 1, one is interacting with the real world, in mode 2 interacting with other people, perhaps indirectly (by reading, for example) and in mode 3 interacting with oneself. Skemp sees discussion as a kind of testing of one's ideas with other people. He points out that mathematics teaching has seemed to rely on a single functioning – communication – chiefly occurring through a single person, the teacher, or through written forms such as workcards or textbooks. Skemp clearly believes that communication should be widened to include talk with other children, and that the other five functionings should be brought into use by them as much as possible. This could happen by the use of mathematical situations based on doing, talking and recording. The doing would bring mode 1 into play, as we have seen in a variety of my extracts. Talking clearly draws the other children into the communication and the discussion (mode 2) – even if you are not entirely happy with Skemp's view that discussion is a kind of testing.

Apparatus is not essential to testing, although it is often helpful. For example, the children in the cube game and the discussion about 'And' are mainly testing one another's ideas, in a manner close to Skemp's meaning, although one group has materials and the other has none. The functioning by internal testing is well illustrated, I think, by the extract 'Trouble With Decimal Places', in which Ann struggles to maintain consistency with her schema of decimal place values. This functioning is likely to be stimulated either by the discussion which is occurring, with other children as well as with the teacher, or by the experimental testing going on. Creativity may occur partly through the use of investigative approaches which emphasise the children's ideas, as we saw in the previous section, and partly by the use of their own forms of recording in the development of symbols.

Skemp (1986) has used his own theory in the design of the materials in his *Primary Mathematics Project* (PMP). I tried to give an early, simplified version when I first introduced building and testing in Chapter 2. It seems to me an illuminating way of thinking about our own teaching of mathematics. It can help when we plan situations, offer insights into how children are thinking, and guide our interventions. I have already made use of its main features - building and testing - in earlier chapters, and will apply them further in my next extracts. But first, I shall bring together some of the main points in planning effective mathematical situations.

Factors in planning effective mathematical situations

In this section I shall try to list and describe briefly the main points in planning our situations which have emerged from the preceding discussions.

1 The situation must stimulate initial interest, and having generated mathematical thinking by the children, continue to sustain it. The activity should continue over extended periods of time - a lesson or several lessons, for example.
2 The activities should involve at least two of DOING, TALKING and RECORDING, often all three. The doing could be interpreted as the confident use of a mathematical idea as if it were an object in the real world.
3 The situation must enable the children to think and work at their own level and make their own judgments about the ideas that come into play. Or, put in Skemp's framework, it should

enable both building and testing of ideas to take place, using some or all of the six functionings he describes.

4 The situation should, as far as possible, enable the children to work on their own, without intervention from the teacher. This point is important, as we shall see in Chapter 7, in organising group work. The game idea that I have illustrated at various stages seems very helpful in this respect, and also in meeting point 1 above. James and Mason use the term 'self-generating activity' for ideas such as the cube game, which keep the children involved for as long as necessary.

5 Try to use a variety of embodiments of a particular concept, rather than just one (James and Mason recommend a minimum of three). Dienes was the original protagonist of this idea of 'multiple embodiment'.

6 Encourage children to develop their own ways of recording their ideas and then introduce the 'official notation' through discussion.

7 Children of all abilities should be able to contribute to a discussion or activity, particularly where a mixed ability group is used. In the case of a game, introduce a chance factor such as dice, or ensure that the outcome of the game is seen as a group achievement. Extra time may be needed by lower ability children to work in the activities, in order to develop meaning (see point 1 above). Children profit from sharing good ideas with one another, so 'streaming' within the class should be treated with care. On the other hand, there are occasions when, because of Cockcroft's 'seven year gap', sharing can have little value for some children.

8 Look for ideas from the mathematics scheme in use in the school (eg from the teachers' books) or which arise from topic work elsewhere in which the children have been, or will be, engaged. Work in other parts of the curriculum such as art and craft may contribute, just as mathematics can be applied in other areas. Also use the language or vocabulary guidance contained in many local authority guidelines.

9 Opportunities for creativity (as in Skemp's mode 3) need to be provided – either as applications in other curriculum areas or investigations. Staff need to be clear about the purposes of these activities, as mentioned earlier. A general feeling of their purpose also needs to be communicated to the children. This will be discussed more fully later in the chapter.

10 Staff benefit from sharing good ideas, so some system of doing this could be set up – this also is discussed in another section.

In listing these factors to be borne in mind, I am not suggesting, of course, that discussion is something that is taking place all the time. There will have to be plenty of opportunities for individual thinking and for practice of skills. But it is clear from the evidence I offered in Part One that there is a serious imbalance to be rectified. I am sceptical about statements made at in-service meetings that discussion will occur 'incidentally'. It is more than likely that the talk will be 'question and answer', 'Guess what I'm thinking' or 'This is how you do it' rather than discussion as I have tried to define it. However, if you experiment in the ways indicated in Part Two you may find that some mathematical situations tend to support discussion and lend themselves to it, while others seem best suited to individual investigation. A rough guide is that situations which offer a variety of opportunities in the building phase are more likely to support group discussion. Others can have talk 'grafted on' to them by devising a suitable game. Problems which only require a single line of attack, or which are cracked by a particular insight, will usually lead to your children working individually until someone thinks of the idea needed.

When you want a group discussion, always ensure that there is ample thinking opportunity for individuals during the building of ideas. *Do not rush* to comparing and testing too early, but allow ideas to develop in a relaxed way. Trialling is really the best guide, and then it is helpful if ideas which have worked well can be shared with other members of staff. If the activity seems particularly effective, why not write it up a little more fully and have it circulated by your local authority? Mid Glamorgan, for example, publishes a termly *Mathematics Box* containing suggestions for starting points.

Sharing ideas among school staff

Exhortations from headteachers or pleas from mathematics coordinators about sharing ideas are likely to fall, not so much on deaf ears as on ears which are far too busy! Discussion over coffee may be good, but the ideas involved tend to be forgotten by the time the opportunity to use them arises. Thus, in my own case, I find myself irritably thinking something like 'That was a good idea that so-and-so was telling me the other week – now what was it?' Either I have to spend time reconstructing it, or find 'so-and-so' to tell me again. An effective way of overcoming this difficulty is to devise a brief but helpful report sheet, so that teachers can record their trials

CLASS	NUMBER IN GROUP	TYPE OF GROUP	TEACHER DIRECTED/ CHILD CENTRED	AREA OF MATHEMATICS
Rec - 4+ yrs	4	same ability	teacher directed	number bonds 5

AIM
1. To produce all number bonds of 5
2. To develop group co-operation

METHOD

2 sets of Unifix eg red/orange
Children close eyes and choose 5 blocks at random
Make own number picture eg ▯▯ (arrange spatially)
Match cubes to card with sticky squares

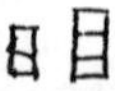

EXAMPLES OF DIALOGUE

Peter I put that first - I've got three red and two orange.
Steven I've got the same as you but I've got three orange and two red.
Jodie I've only got one red and I've got four orange.
Charlene I've got the same as Peter.
Steven Look I've got all red this time so now all we need is all orange and then we've finished.

The activity was repeated to find the missing number stories.

COMMENTS

Although the activity was initially teacher-directed very little contribution was made by me. The children carried on until the task was completed. It took four attempts until someone had 5 orange.

The situation promoted discussion and group co-operation, at the same time they were learning the number bonds of 5.

LANGUAGE (VOCABULARY) - number names one to five, colours red/orange, same as, different, more, more than, less than, less.

FIGURE 6.4

of mathematical situations (of course, the idea could work just as well for other parts of the curriculum). Copies of this could then be available for other teachers to consult and use at an appropriate point in the scheme of work. The reports can also be used as a basis of a staff in-service or development discussion. A possible format for such a report sheet is shown in Figure 6.4, along with an actual report. The details include the kinds of language that children used in the course of the activity. You may find this approach one that you can adapt for use in your school. It will be important, of course, not to let the outcomes in the report determine what happens when the activity is repeated - that would destroy discussion!

A group of teachers in Rochdale has prepared a format for similar 'desk top guides' to mathematical activities, although these are not cast in the form of reporting back on trials, which the Lakatos Group had in mind in devising the form shown in Figure 6.4. The Rochdale format is no larger than two sheets of A4 paper, and covers information about setting up the activity (such as resources needed), objectives covered, ideas for teaching and space for the

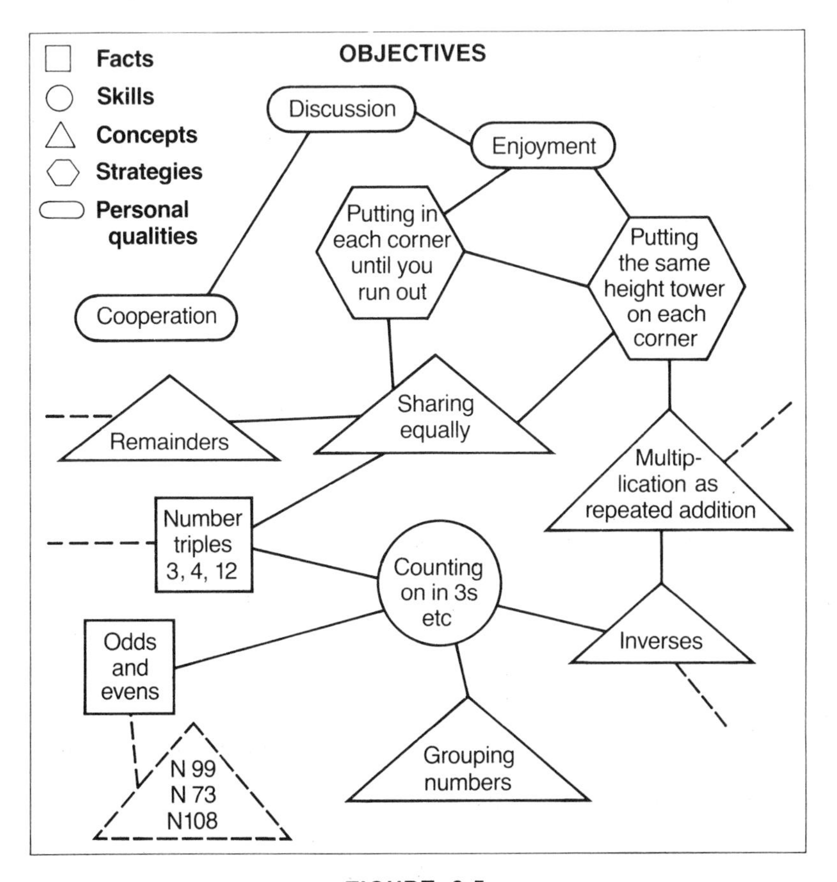

FIGURE 6.5

teacher's personal notes. The format for the objectives is particularly ingenious, using a distinctive shape for each of the five categories listed in *Mathematics 5–16*, as indicated in Figure 6.5. for the *Tower Game* described in *PrIME Newsletter* No 3 (1987). If an activity seems particularly suitable following staff trials, it could be embodied in this more detailed form as a 'desk top guide' by the mathematics coordinator.

In making such records, it is helpful to other teachers to include representative 'key comments', and also the kinds of language that the teacher may inject into the situation to negotiate meaning and develop the children's use of language. This is, of course, an ongoing process throughout the primary school years, in which teachers gradually enlarge the 'informal language' in discussion and recording with a 'formal' mathematical vocabulary. I have already described the views of James and Mason, and of Skemp, on how this might be done. We need to bear in mind that early meanings of mathematical language and symbols are very often radically developed in later learning. These early meanings need to be flexible enough, and linked with sufficient experience, to allow later developments to take place with understanding. Besides the 'desk top guides' to particular activities, 'maps' showing the relationships between the ideas in activities could also be helpful, to help

	Informal language ⟶	**Formal language**
Number	add, and, total take away, more/less than times, timesed by, by, share, shared by makes, is the same as, gives	add, addition, sum, +, plus minus, subtract, difference, −, subtraction multiply, multiplication, ×, product of divide, division, ÷, over equals, is equal to, =
Space	corner amount of turn oblong, square diamond wheel, round, lid, spoke	vertex angle, rotation, degree rectangle rhombus circle, centre, radius
Logic	idea, guess because works, is right/wrong if....then pattern (in real contexts)	conjecture reason verify, prove/disprove implies, follows from, ⇒, ⇐ ⇒ pattern (in symbols), sequence, series, tessellation

FIGURE 6.6

teachers plan the progress of groups through the work over a period of time. I shall include examples in the next sections, in relation to division and to whole number properties.

There needs to be guidance in the mathematics scheme, on the lines asked for by David Lea in Chapter 1, about the stages at which mathematical language should be introduced, and the kinds of experience which should underlie it. An authoritative guide is beyond my scope, but the guidelines issued by many local authorities could provide a helpful framework in this task. Sometimes these use the term 'vocabulary' rather than 'language' in their descriptions. Some examples of possible developments of meaning that can take place are suggested in Figure 6.6.

Using the equaliser balance

This piece of apparatus features in many primary schemes, such as those of Nuffield Mathematics or the Scottish Primary Mathematics Group. Its use links with experience of using balance scales in weighing activities. By hanging identical metal rings on equally-spaced hooks or (in this case) placing weights in slots, numerical relationships emerge. The balance was designed by Dienes to act as an alternative embodiment of multiplication as repeated addition.

A group of three average ability top infants were asked the general question 'What do you think this is for?' The aim was to encourage a problem-solving approach to the use of the equaliser, and to develop recordings to link with other experience using rods.

Andrew: It's a number scales.
Lissa: It's for weighing.
Nicola: I'm not sure, is it for balancing things?
Andrew: It's got slots to put these bits of metal in.
Lissa: I think these weights are different - some are light, and some are heavy.
Andrew: I don't think so! They're all the same weight, let's test it.

(After some further exploratory work the children found that the beam either balanced evenly or swung down. They agreed unanimously that the weights were identical, their positioning determined what happened.)

Lissa: Miss, there's four weights on each side, but that side adds up to twenty-four and the other side is thirty-one. It's not balanced but thirty-one is a higher number, it's seven more than twenty-four.

Andrew: Miss, if I can put two weights in a ten slot that will be ten and ten making twenty. I can make twenty on the other side with three sixes and a two.

Lissa: I can make eight with four and four on one side and six and two on the other side.

Nicola: It's my turn. I'll make twelve with two weights in the six because two lots of six makes twelve and three weights in four because three lots of four makes twelve.

The children were quick to experiment with other weights; though mistakes were made, discussion and activity continued enthusiastically. Judging from the comments such as 'It's my turn' the children were able to cooperate well and talk together. The teacher's questions at some appropriate point can ensure that the links with addition and multiplication become explicit – it is clear from their comments that the children are already doing this implicitly. A good way is to make them predict – this can be used a lot and the children may learn to use the technique themselves. For instance: 'I'm putting a weight in the ten slot. I want you to think of a way of balancing it with two weights – don't say anything yet'. Children are called on to make their prediction, and the others to agree or disagree – this will ensure that everyone has a stake in the outcome when testing takes place. Notice once again how the equaliser balance contributes to BUILDING and TESTING. Patterns will emerge from these predictions and the children can make recordings in their own way, and decide whether to use numerals. Records such as $4 + 4 + 4 = 12$, $5 + 5 = 10$ can then be linked with $3 \times 4 = 12$, $2 \times 5 = 10$ etc, depending on how the children decide to present their findings. We can infer that most, if not all, of Skemp's functionings are operating in this situation.

In order to keep the activity going for longer, and help the children to build their ideas fully, it would be a good idea to devise a follow-up game using the equaliser. For example, two dice are rolled to decide where to place a weight on the balance, then children take turns to make predictions, as above. The idea of 'predict' then 'challenge or accept' that I have demonstrated in previous accounts might be used here, and recordings could be incorporated in some form. This would help the teacher to follow up what had taken place in such an activity when she was not present. The above account is clearly to do with the 'conceptual structures' category of aims and so comes towards the closed end of my dimension (see page 115) – the goal of this activity is set by the teacher. Appropriate interventions are made to ensure that its purpose is achieved and to make links with other relevant experiences - it would be optimistic to expect children always to do this. The introduction of a game based on the activity will counteract the oft-observed tendency for children to hurry through, for obvious reasons. My next account, in contrast, is an exploratory activity towards the open end of the dimension. The goals are being set by the children, within the general constraints planned by the teacher.

Exploring with Clixi apparatus

Clixi is a useful structural apparatus which affords scope for free play investigations by the children, in which quite complex structures can be devised. Ample opportunity arises for language development, for example naming the shapes, talking about simple properties, using relations such as 'next to', 'underneath', 'above', 'joined to' etc. Formal terminology may include terms like 'edge', 'face', 'cube', 'prism', which need to be made meaningful through DOING and TALKING, as usual. The teacher can also assess informally, through her observations and conversations, qualities such as imagination and persistence which are listed under the 'Personal qualities' category of aims, and encourage children appropriately. Creativity is clearly the main functioning being encouraged. Some examples of actual dialogue follow.

Gareth: Miss, I've made a cube.
Teacher: That's good, Gareth. How many sides has your shape?
Gareth: Miss, I've got six sides, all square shapes.

The teacher could continue to leave the direction of the investigation to Gareth, or shape it herself with a probing question such as 'I wonder if you could make me a larger cube, Gareth?' In the latter case, the teacher takes control of what is happening, and the activity moves towards the closed end of my dimension. Gareth may lose interest or may be challenged to build his ideas further. It may be better to challenge in this way on a later occasion, and retain the distinctively open character of the present activity.

Barrie: I can make a cuboid if I join two squares to make a rectangle.
Teacher: How many rectangle shapes are you going to need for your cuboid?

Barrie: Four, Miss, one for the top, one for the bottom and one for each side. For the ends I need squares.

Teacher: How many pieces of Clixi have you used altogether then?

Barrie: Um . . . ten square pieces Miss.

(Here the teacher intervenes just enough to find out about Barrie's thinking, without taking over the direction from him.)

Lisa: Miss, look what I've made!

Teacher: That's interesting – go to the solid shape display, Lisa, and see if you can find the name of your shape.

Lisa (later): Miss, I know – it's a square pyramid!

Teacher: What shapes did you use?

Lisa: Miss, a square and four triangles coming together to a point at the top.

Jessica: If I put these shapes together flat on the table edge to edge I can make a lovely pattern, like our shape patterns on the wall.

These comments are typical of the discussion which flows when children use *Clixi* in a free play situation, both between the children and between the children and the teacher. Other types of structural apparatus can be used in a similar exploratory manner. The experience will feed into more directed activities, which are towards the closed end of my dimension, at a later stage. The school

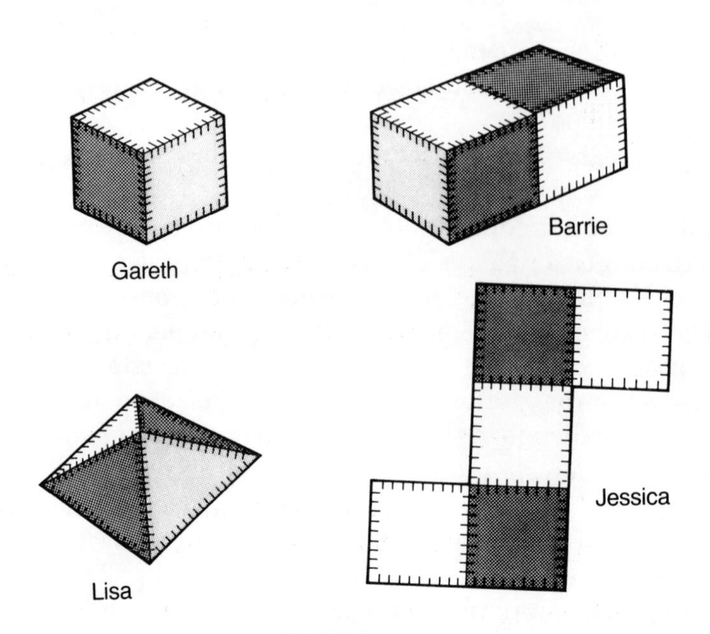

FIGURE 6.7

mathematics coordinator should (as here) make records which can offer guidance to other staff on the kinds of language which may emerge, and the experiences which will help in later understanding of mathematical ideas. Apparatus such as *Clixi* might be used with one small group, or with different groups involved in varieties of activity. Recordings could be made, if necessary, by drawing, writing or photography, but most simply by display of the actual results. The shapes made by Gareth, Barrie, Lisa and Jessica are sketched in Figure 6.7.

Unifix cubes and sums

A group of older infants were given a set of *Unifix* cubes each and asked to make up sums, talking to and helping each other as they worked. The aim was to stimulate language development in number concepts, around the set of twenty cubes chosen by the teacher. The activity was related to the current stage of the Nuffield scheme used in the school. Clearly we are once gain towards the closed end of the discussion dimension, with specific goals set by the teacher. Typical dialogue was as follows.

Vicki: Miss, I can put two add two add two add two . . . add two . . . (She gets lost.)

Ann: Look, five add five add five add five make twenty. (To Joan.) Does twelve add nine make twenty?

Joan: No! Twelve add nine makes twenty-one.

Teacher (after a short while): Can you give me the signs when you are working out sums?

(Several): Yes!

Teacher: Is that sum correct, Ann? (Ann has written 11 + 9 = 20.)

Ann: Yes, Miss, because I've got it down on my paper.

Emma: And I've got it.

Joan: And me!

Lionel: And me!

Teacher: So you agree then, that it must be right because you've all got the same answer?

Children: Yes.

(Joan has written 21 − 20 = 1, 21 − 1 = 20.)

Teacher: Vicki, can you check if that is correct?

Joan (butting in to Vicki): Here's a spare block, now you have twenty-one. Take it away and you'll have twenty left.
Teacher: What are those four cubes for, Darcy?
(Speaking to Darcy, who has been working out division sums.)
Darcy: Miss, there's invisible people by them, there and there and there and there (pointing) – I'm sharing them out.
(Emma has written 20 – 10 = 2.)
Teacher: Why is that, Emma?
Emma: Because one ten is ten and two tens are twenty.

These fragments illustrate how discussion enables the teacher, by listening and observing, to diagnose misunderstandings as well as understanding. Thus she or he will often have to make on-the-spot decisions as to how to intervene. What Emma has just said appears to make sense, but her recording of her idea does not – although presumably it does to Emma. A very useful all-purpose resource is the 'Show me – ' request, which you have seen in action in earlier accounts (eg in 'Trouble with decimal places'). Here the *Unifix* materials are available to help, and so the teacher might ask 'Show me what you mean with the cubes when you said two tens and twenty'. Later, perhaps, the teacher might write down a sum such as 10 – 4 and ask Emma again to show what it meant with the cubes. Using such information the teacher could re-negotiate the

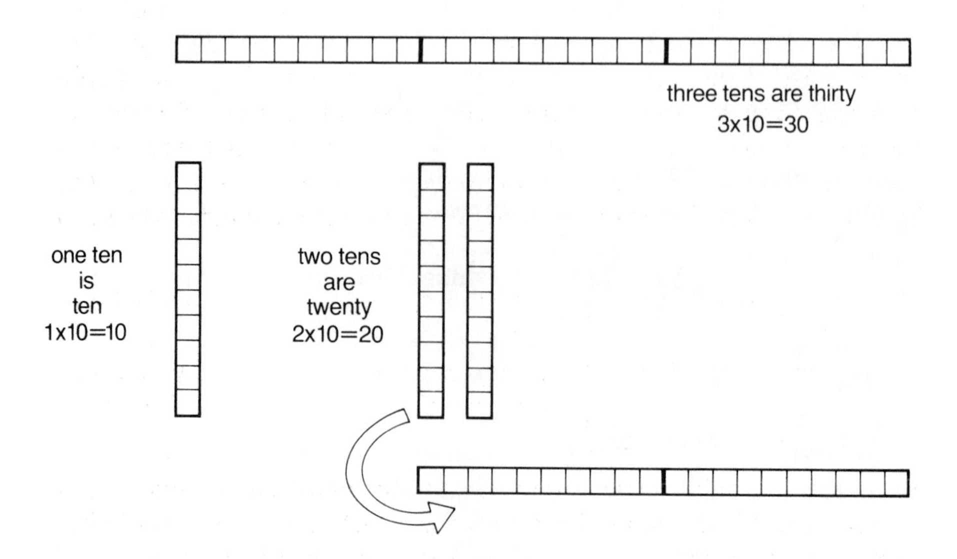

FIGURE 6.8

meaning of Emma's recordings with her. This process can be based on the informal recordings idea of James and Mason which I described earlier. Emma draws her shapes and the teacher helps her to write what she says. Thus she moves more slowly from the enactive stage of making her shapes, though informal iconic records and talk, to using the symbols, which in her case seem poorly understood. Diagrams such as those in Figure 6.8 might result, and be linked with other embodiments, such as counters. Other children such as Joan might be drawn in to help, by saying what they think is meant by the same questions. Vicki also seems to need help - how might one intervene in her case?

Subtraction needs to be treated as both 'taking away' and 'comparing'; 'counting on' is yet another important form. Children should experience suitable embodiments and situations involving all three, so that they do not develop an unduly narrow meaning focused on 'take away'. A point worth noting is that a narrowly-based conception arising from early practical activity may become a misconception later on, or prove unhelpful in the development of later concepts. For example, the association of subtraction with 'taking away', as evidenced by Barrie and Gavin earlier and by Joan in the previous extract, if it is identified with the meaning 'take smaller from larger', could become a misconception when moving on to problems such as 25 – 17, by suggesting that you perform 'two take away one' and 'seven take away five'. Problems over subtraction are also experienced by the group discussing 'What's the difference?' in another extract. It looks as if the form of symbolic recording using the subtraction sign should be linked with a wider range of language than 'taking away' - the language that the children I encounter invariably seem to use. Perhaps we should postpone the symbolic form until a suitable range of informal recordings has been developed, related to specific contexts on the lines suggested by James and Mason, and then introduce the formal notation along with a 'neutral' form of spoken language such as 'minus', eg

'Take away' experience 'Comparing' experience 'Counting on' experience	}	informal recordings using 'context language'	⟶ symbolic records (spoken as 'minus')

Proceeding in this way would mean that the children in the previous extract would be using their own forms of recording - probably iconic - at this stage rather than moving directly to the use of symbols and operations as reported. Similar stages of development would be used in each of the various forms of subtraction experience, as suggested in Figure 6.9. The teacher moves by stages towards the final symbols, using the children's own forms of recording based on their manipulations of objects. The children are encouraged to talk and discuss with the teacher and among themselves, so as to establish a wide basis of meanings for the symbols when they are introduced.

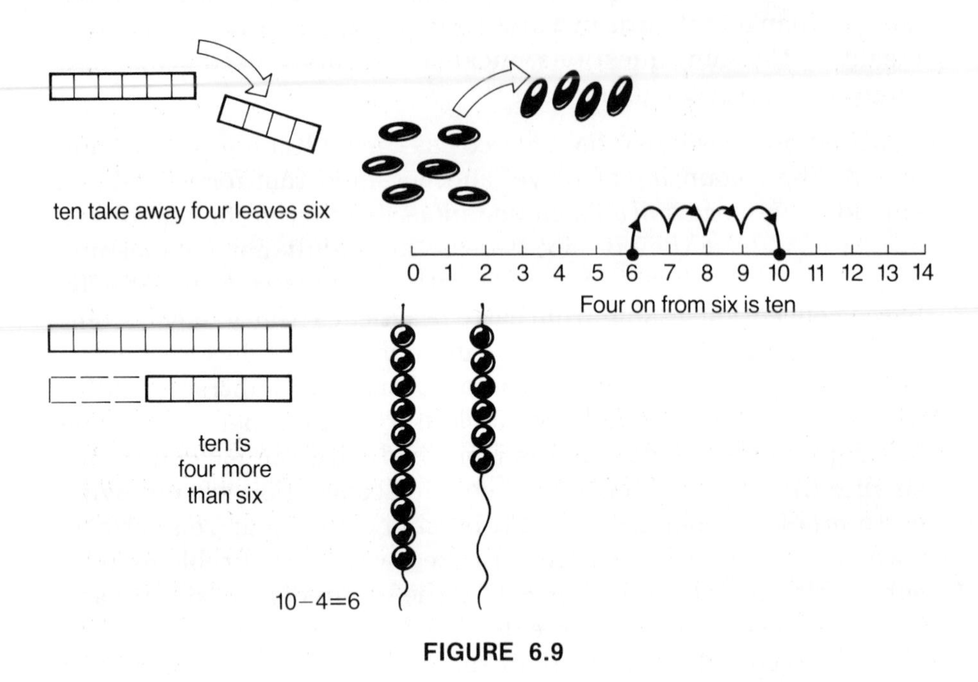

FIGURE 6.9

Yet another development over subtraction is foreshadowed in the calculator game near the end of Chapter 5, where negative numbers appear. Here the number line being used by Craig and Steven for subtraction may be used in another way so that 1 – 7 becomes possible after all, and yields another meaning, namely – 6. When do children start 'owing one another' or 'lending to/borrowing from one another'? The infants 'In the shop' in Chapter 4 show no trace of it, but certainly by the later junior stage this type of experience with money seems to be available to the children. This provides a possible embodiment for the idea of positive and negative numbers, as well as for how they should be 'added'. Another available experience might be the weather charts on television which show temperatures above and below zero in yellow or blue rings. But the 'take smaller from larger' idea is a block to this development, whichever model is used.

Other interesting information emerges from the talk – for example Darcy's comment which shows him imagining four people present to help him in his sharing, with the aid of his four cubes. What do you think of the children's understanding of testing, after the teacher asked 'Is that sum correct, Ann?' One of the things that the teacher might hope is that, after periods of such discussion, the children will begin to use questions such as 'Show me . . .', 'Why?', 'What do you mean . . .?' and so on, to one another.

Difficulties with division are similar to those with subtraction. Children invited to write stories about division seem, like Darcy, to fasten on to sharing, and the language 'shared by' can also appear to denote the operation. This is associated with another misconception, that division is always of the larger by the smaller number. There is another type of experience associated with division of course, that of grouping. Darcy is sharing out his twenty cubes among four people so that they each get five, as in Figure 6.10(*i*), but could instead count out sets of four to find that he ended with five sets, as in Figure 6.10(*ii*). In the past, two distinct signs were used to distinguish these two senses, 20 ÷ 4 to denote sharing and 20:4 to denote grouping. An appropriate story for the latter might be 'John has eight bananas and eats one for lunch and one for tea each day – how many days will they last?' But 'sharing language' is still likely to be used because no real alternative is available to children, as far as I can see. Or the idea of division is not really linked with the grouping experience, because multiplication can be used instead. Here are some infants, for example, asked to find how many squares they could build using twelve lengths of straw.

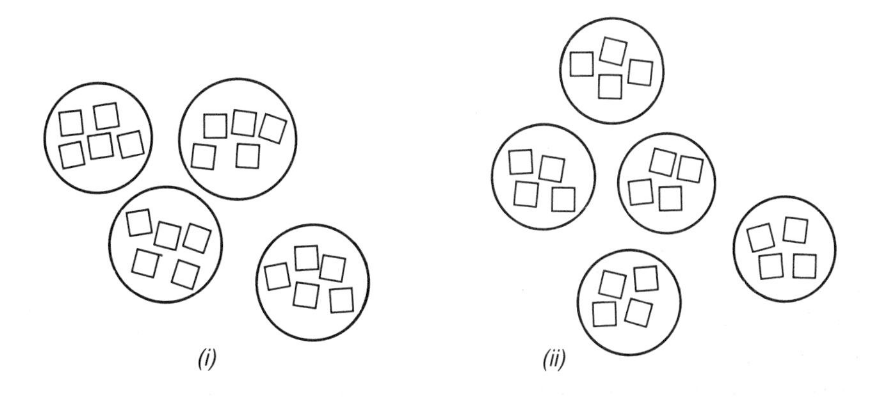

FIGURE 6.10

Barrie: Miss, share the straws to make four equal groups because you need four to make one square and – um –

Jessica: No Barrie, wait a minute I know – Miss you need four to make one square so take four away then another four and so on until there's none left.

Gareth: A square has four sides so it takes four straws to make one square, then four add four makes eight, it takes eight to make two squares and another four will be twelve straws used and we can make another square. That's one, two, three squares.

Gavin: Miss, see how many fours in twelve, because you're going to need four to make one square, there's three lots of four in twelve, so you'd make, um, three squares, I think.

Nara: We could write it down like before in equal grouping Miss, twelve equals three times four or three times four equals twelve, it's all the same.

The important survey of the Assessment of Performance Unit findings made by Quadling *et al* (1985) discusses 'Number, Language and Context' in chapter 3. It quotes the chocolate bar problem in which a chocolate bar can be broken into 12 squares. There are 3 squares in a row, and children are asked 'How do you work out how many rows there are?' 34% gave 3×4, 32% gave $12 \div 3$ and 13% gave $3 \div 12$! This supports the assertion which I made above. The survey says that modern textbook writers make a conscious effort to include a great variety of ways of expressing number operations – but often in the fewest pages, it seems! Time and talk are what is really needed, of course. Chapter 3 of the survey also discusses the calculation techniques used by children. No fewer than eight *common* methods were used in response to a question about how many weeks it took to save up £4.50 for a record, saving 25 pence each week. In the past a variety of algorithms was employed until the present 'long division' form emerged as the most commonly-taught. This is clearly related to the grouping experience, since it is based on successive subtractions of multiples of the divisor from the dividend. So a technique based on grouping might be applied to a situation involving sharing, eg share 28 sweets among 6 children. The survival of this hallowed technique is now very much in question, as we saw from an earlier quotation from *Mathematics 5–16*.

Practical applications of the division operation which are to do with grouping rather than sharing must be puzzling to children whose ideas about it are context-bound to the latter. But many later applications of division in mathematics seem not to have an obvious relation to either sharing or grouping. For example, if I travel 150 kilometres in my car and take 3½ hours, what is my average speed? Or, if a rectangle has an area of 100 square metres and a length of 15 metres, how can one calculate its breadth? We have left behind the situations in which only whole numbers arise, and are in those which allow decimals or fractions to be multiplied. Here, the understanding required seems to be more formal, that division is an 'inverse operation' which 'undoes multiplication'. Very often a fraction bar will be used rather than the division sign, along with language like 'distance over time equals speed'. Previously the children may have met the fraction bar only in connection with language such as 'three quarters' and ⅝, so here is another source of possible confusion unless careful teaching takes place. This difficulty happens when mathematics teachers introduce formulae such as

$$\text{distance} = \text{speed} \times \text{time} \Rightarrow \text{distance} \div \text{time} = \text{speed}$$

or $$\text{area} = \text{length} \times \text{breadth} \Rightarrow \text{area} \div \text{length} = \text{breadth}$$

If calculating the result of a division operation becomes a matter of button-pressing on a calculator, the situation will resemble that with square roots as in my account in Chapter 5, 'The calculator as a Black Box'. The children there, like most people, have no idea how to calculate the square root of 57, but they understand, I think, what the calculator is doing for them when it produces the decimal. A similar situation will need to appertain with division in the future. However, 'square rooting' is a 'unary operation' in which you operate on one number only, while division is a 'binary operation' which operates with two numbers. Unlike multiplication, it is vitally important which is the divisor and which the dividend. I suggest the use of black box calculator games to help develop this more formal meaning. The game I described at the end of Chapter 5 could be extended in the following way:

The first player hides a whole number in the calculator memory, and the second player enters another whole number on the display as before. The first player now secretly either multiplies or divides the display number by the memory number, using the memory recall (MR) button and passes the result to the second player. He or she uses the information available, and another calculator, to try to find the hidden number. For example:
First player hides 24 in memory, second player enters 78. First player chooses 78 × 24 giving 1872, second player infers multiplication from 1872, performs 1872 ÷ 78 on his own calculator to get 24.

Now second player hides 35, first player enters 15, second player chooses 15 ÷ 35 = 0.428 . . . The first player should now calculate 15 ÷ 0.428 . . . Why?

The game could be played for fun, or to score points as in squash – failure to get the hidden number gains the other player a point, and the hidden number 'changes sides' when found. This game can become very subtle if extensions of the rules are allowed, such as use of integers or decimals as input, or using the other operations. You can see that the choice of division rather than multiplication forces some clear thinking from the opponent. In the previous example, the first player can infer division has been used, since his entry of 15 has got smaller and whole numbers are in use. So

$$0.428\ldots \times \text{'hidden number'} = 15,$$

and thus the division mentioned above will find the hidden value.

The survey by Quadling described above contains an extended discussion of these important matters in chapter 8, on 'Algebra'. It suggests that the computer or calculator could provide a means of giving concrete meaning to algebra, making it more attractive to today's generation. 'Whatever approach is used, it is likely that this "meaning" will only be acquired through verbal exchanges carried on over an extended period of time.'

My discussion of subtraction and division illustrates how the longer term developments of meaning have to be borne in mind by

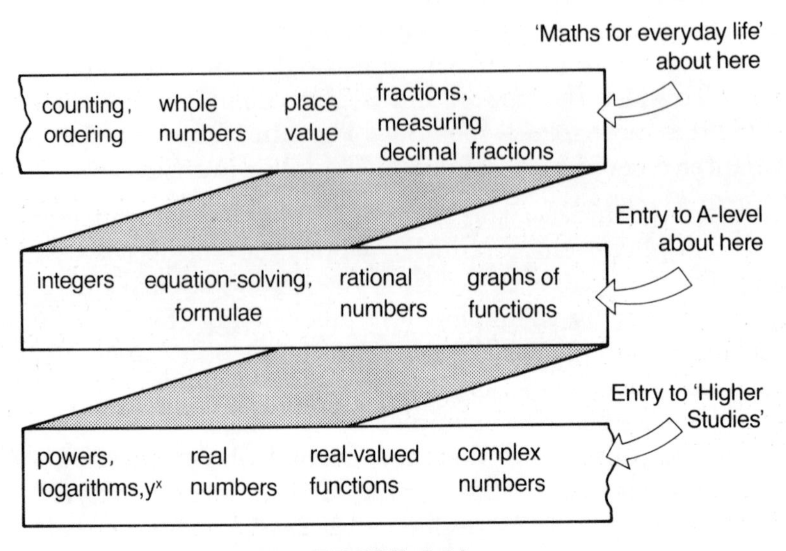

FIGURE 6.11

the teacher as she or he plans the early learning experiences, the point I raised earlier. There are constant further developments, so that early meanings will have to be extended and re-negotiated. It might be helpful for primary staff to have a general idea of 'where the number system goes' later on, somewhat as in Figure 6.11, but in rather more detail.

Describing unseen objects

Situations based around the idea of children describing objects which are unseen by others, for them to identify or draw, are a very powerful way of developing language and observational skills. We have seen this idea at work already in the extract 'In the Touching Corner' in Chapter 3. Variations of this idea can be found in many sources, and the approach can be used in other parts of the curriculum than mathematics. Thus Needham (1984) describes a 'Back to back' game, in which two children make matching sets of coloured shapes. Sitting back to back, one of them arranges his or her set to form a pattern. This child then instructs the partner (who cannot ask questions) step by step, trying to create an identical design. When the instruction process is finished, the children turn and compare their designs. This nice idea obviously might have mathematical possibilities. A similar situation, quite well-known, involves drawing a network which has to be described and reproduced without actually being seen. The two following ideas have worked successfully with various children.

1 Guessing Logiblocks

This activity is for two to four children at the infant stage. One member of the group sits behind a screen with a supply of *Logiblocks*. He or she chooses one and the others have to ask questions requiring 'yes' or 'no' answers, in order to guess what it is. The activity flowed well with older infants, with the aim of encouraging them to ask questions such as 'Is it round?', 'Is it thin?' so as to eliminate various sets of blocks. Interestingly, with younger infants (four to five years) the activity foundered because the children could not ask appropriate questions, although when the teacher took the guessing role, they could respond appropriately.

2 Guessing or drawing a hidden object

This activity could be used either with a whole class, divided into small groups of about four, or with one group only. A set of mathematical objects is placed in a large carboard box – they could be objects about which the children have recently been learning such as a cube, cuboid, pyramid or cylinder, or everyday objects, pictures or shapes. One child is chosen to come to the front and draw out an object, keeping it concealed behind a flap. She or he describes it, but is not allowed, of course, to name it, or mention its use if this applies. The group(s) discuss and try to identify the object – either naming it or drawing it. Groups or individuals ask questions in turn (training in turn-taking!). After a certain number of questions, or successful identification, the object is revealed. The game may be played competitively, or just for fun. The skill of trying to draw various objects is well worth cultivating, although this form of the activity is probably suited to older age groups. A simpler version is to get the groups to identify which object is being described from a set displayed in different places around the room. Describing, discussing, listening and observational skills are then all being exercised.

This type of activity probably falls somewhere midway between the 'open' and 'closed' dimensions of Figure 6.1. A very open mathematical situation may lead to more closed follow-up ones, or vice-versa. For example, the cube game of Chapter 3, and the drawings using triangular spotty paper, are fairly closed – the teacher sets the goals. The rash of isometric drawings that subsequently caught on in the class was very open and creative, however – sparked off by the ideas generated in this small group. Shading and colouring, and even the idea of 'impossible objects' lead in the direction of art and craft work. The reverse process is illustrated by an activity in which children (upper juniors) were invited to make up shapes and patterns, and colour them, using triangular grid paper, or later, squared paper. The children showed considerable imagination with the creation of abstract shapes and patterns and also named things, for instance dragons, dinosaurs, tanks, pyramids, tents, crocodiles and even (on one never-to-be-forgotten occasion) 'grandad's false teeth'! The shapes and patterns included all those needed in geometry – equilateral triangles, hexagons, rhombi, trapezia and so on – to be drawn out in more specific discussions later on. Squared paper, on the other hand, may generate buildings and robots. Two rather more closed activities with the same materials will now be described.

Polyominoes

This idea is sufficiently well-known not to need detailed description, but it lends itself to small group discussion. The children are asked to find and draw on squared paper all the shapes containing three squares, four squares, five squares and so on up to seven. The investigation gets harder, of course, and soon generates 'contributory discussion' – 'Have you found this one?', 'I've got four – how many have you got?' Rather later, 'comparative discussion' gets going – 'Those are the same – that one doesn't count', 'They're not the same!' and so on. This type of argument centres around whether mirror images or rotations count as different. The situation changes when the shapes are cut out and can be turned over or round – as with the 'pentominoes' (made of five squares) in Figure 6.12. Follow-up activities include making tessellations and rectangle puzzles or deciding which pentominoes or hexominoes will fold into cubes. Teacher intervention can be minimal, perhaps enquiring whether any system is being used in finding the full set of the later polyominoes.

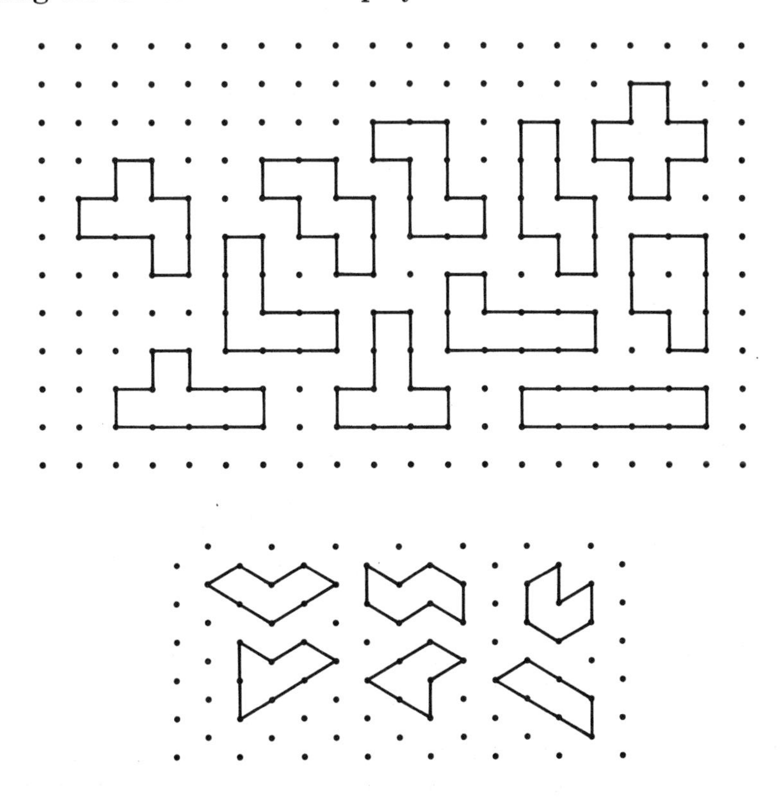

FIGURE 6.12

Somewhat less well-known, perhaps, is the extension of this activity to triangular grids, rather than squares. A few of these shapes are also shown in Figure 6.12, and similar types of follow-up are clearly possible.

Rectangular and square numbers

This situation is suitable for older junior children and can be used with the whole class divided into small groups. Each group has a large (A1) sheet of centimetre-squared paper, and a supply of coloured pencils or crayons. The teacher starts off the activity by asking the children to draw as many rectangles as they can which contain 24 squares. After a suitable interval of activity, monitored by the teacher, the group outcomes are compared and discussed. The main question likely to arise here is whether an 8×3 rectangle is to be counted as distinct from a 3×8 rectangle (drawn in different directions on the squared paper, of course). The children are usually keen to count them as distinct, so as to make more shapes. The final pattern which emerges (see Figure 6.14) also looks more complete and symmetrical if this happens. The teacher can encourage 'keeping track of what we've found by using numbers'; this helps to produce records like $3 \times 8 = 24, 8 \times 3 = 24, 2 \times 12 = 24, 12 \times 2 = 24$ etc. The groups are then set a more extended task, of doing the same thing for each of the numbers from 1 up to 36 (say). A question that is sure to arise during this activity is 'Do squares count?' The teacher agrees that they should be included - a class activity on the further point that will arise, 'Is a square a rectangle?', is given in Chapter 8.

The aim of this activity is to investigate the properties of different kinds of whole numbers. The children find that some numbers will make a square, some will make rectangles, but others will only make two 'special rectangles'. Thus 16 will make a square as well as some rectangles, 24 makes rectangles but not a square, while 7 will only make 7×1 and 1×7 shapes. To shorten the time needed for the investigation, the class could be sorted into four working parties, each to work on a set of nine numbers allocated to them. A possible allocation which ensures that each group gets a set containing both large and small numbers of all three kinds, rectangular, square and prime, is shown below.

This gives each group plenty to do and discuss, and the outcomes can be coloured and shared in a final display. This can form the basis of a whole class discussion, in which the terms 'square', 'prime' and

Working party

A	B	C	D
1	2	3	5
4	8	10	12
6	11	13	17
7	15	18	19
9	16	23	20
14	22	26	25
21	24	33	27
28	29	35	30
31	32	36	34

'rectangular' can be introduced, and the corresponding sets are listed:

Square numbers = {4, 9, 16, 25, 36}

Prime numbers = {2, 3, 5, 7, 11, 13, 17, 19, 23, 29, 31}

Rectangular numbers = {6, 8, 10, 12, 14, 15, 18, 20, 21, 22, 24, 26, 27, 28, 30, 32, 33, 34, 35}

The case of 1 can be covered – is it somehow special, or is it the simplest square number? Every number is linked with a set of shapes to give the basis of the 'doing' in this activity, and to enable the testing to be carried out.

There is often an early tendency to confuse the set of primes with the set of odd numbers. Children soon point out that 9 breaks the sequence, since it can 'make a square 3 × 3'. A set diagram like that in Figure 6.13 could be drawn to highlight this relationship between odd, even and prime. A possible follow-up activity for more able children might be to use the well-known 'Eratosthenes sieve' method in a number square up to 100 to extend the set of primes further, and show how it 'thins out' as we progress into the sequence of whole numbers.

It is worth bearing in mind that the children are investigating a set which has 'no pattern', probably for the first time – their previous encounters will have been with number sequences or table patterns, as in my *Ergo* extracts in Chapter 5. The primes have been a source of fascination, as well as many difficult problems, for mathematic-

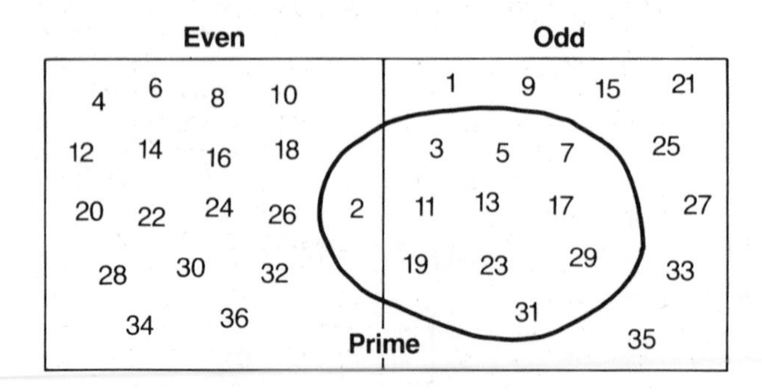

Figure 6.13 Set diagram for primes up to 36

ians in various civilisations for many centuries. They represent the aesthetic, 'fun' side of number work, which otherwise is concerned with applications in industry, commerce, science or technology. It must have been difficult for earlier mathematicians to carry out their calculations, lacking that 'brilliantly simple notation' which we employ. Various devices and models were used, including the ideas of triangular, rectangular and square numbers which we have inherited from them. Multicultural studies could be made showing the contributions of different cultures to our ideas on number, as well as the (rather more common) studies of notation. This 'pure mathematics of number' could form a basis for several types of activity. Groups could carry out number investigations, perhaps using calculators or even microcomputer programs as in Chapter 5 (*Digital roots*). A simple Basic program was described in the *PrIME Newsletter* No 3 (1987) and an emendation enables it to be used by children who know what a prime number is to investigate the number of their birth year.

```
10 INPUT Y
20 INPUT N
30 PRINT Y/N      * NOTE HOW WE TELL THE
                  COMPUTER TO DIVIDE!
40 GOTO 10        ANOTHER DISCUSSION THEME?
```

The child first inputs its year of birth Y, then a trial factor N. The program divides Y by N and prints the result, then returns to the start of the program and waits for another trial number N to be input.

The activity on rectangular, square and prime numbers can be

12	12	24	36	48	60	72	84	96	108	120	132	144
11	11	22	33	44	55	66	77	88	99	110	121	132
10	10	20	30	40	50	60	70	80	90	100	110	120
9	9	18	27	36	45	54	63	72	81	90	99	108
8	8	16	24	32	40	48	56	64	72	80	88	96
7	7	14	21	28	35	42	49	56	63	70	77	84
6	6	12	18	24	30	36	42	48	54	60	66	72
5	5	10	15	20	25	30	35	40	45	50	55	60
4	4	8	12	16	20	24	28	32	36	40	44	48
3	3	6	9	12	15	18	21	24	27	30	33	36
2	2	4	6	8	10	12	14	16	18	20	22	24
1	X	2	3	4	5	6	7	8	9	10	11	12
	1	2	3	4	5	6	7	8	9	10	11	12

FIGURE 6.14

rounded off by a further one involving the construction of Figure 6.14. (This forms one of the later activities in the Nuffield scheme.) The children construct a 'multiplication square' up to 12 × 12 on squared paper. Working in small groups, they are asked to find and investigate the patterns of square, rectangular and prime numbers that have already been identified in a different context using squared paper. In Figure 6.14 the shapes are still present, but overlapping and so less obvious. Usually, the children soon notice the square numbers, and observe that these are at the corners of overlapping squares which run along the diagonal. Other questions that might be asked are 'Is there a pattern for 24?', 'Whereabouts do you find the prime numbers?' and so on. These patterns can be coloured and highlighted by the children as indicated in Figure 6.14. The way in which the set of square numbers can be continued will be apparent, and some children may pursue this. Another follow-up might be in the form of a game, as described by Richard Skemp in his *Primary Mathematics Project*. The children have a set of counters, select a subset and take turns at trying to construct fresh rectangles from it. Played with an initial set of 36, such a game would certainly reinforce the ideas built into the activities I have described and provide practice in applying them.

The mathematics coordinator in a school will probably need to indicate the mathematical richness of these situations to other staff, as possible sources of further work. These activities are helpful if interposed between introductory work on area, such as tessellations, covering and measuring leaves or hands etc, and later follow-up. You will be very unlucky if children cannot quickly find a rule for finding the area of rectangles of the form

Number in a row × Number of rows

The leap from this whole number situation to that where any numbers are multiplied (or divided) could be made by making use of the pattern of 24 in Figure 6.14. This activity would be more appropriate for more able children or those of early secondary school age. On squared paper the rectangles with an area of 24 are shown overlapping, as in Figure 6.15, and then a smooth graph is drawn through the vertices of the rectangles as shown. This curve can be used to investigate other rectangles with an area of 24 – for example such a rectangle which has one side 7 units long will have its other side fixed by the curve, and this length can be measured

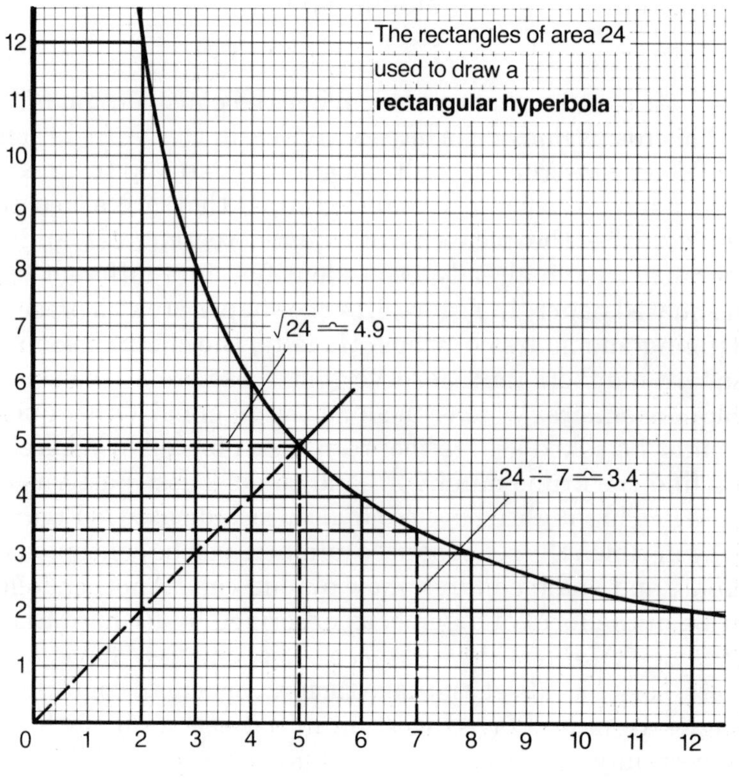

FIGURE 6.15

and then checked by calculator, as 24 ÷ 7. This rectangle is drawn with broken lines in the figure. Is there a square with an area of 24 units? How long is its side? This approach gives another way of looking at square roots, and could be linked with the corresponding calculator activity in Chapter 5. In this way the square number investigation could be pursued even further. So the curves which form patterns in Figure 6.15 (for other numbers as well as 24, of course) could be used to extend the rule for the area of a rectangle more meaningfully from whole numbers to decimals, to develop the relationship between the multiplication and division from sharing and grouping, or for square roots and square numbers. The technical name for this useful curve is 'rectangular hyperbola'.

I mentioned previously that a 'map' showing the relationships between activities in or around a topic could be helpful in planning progression. The map for this particular group is shown in Figure 6.16.

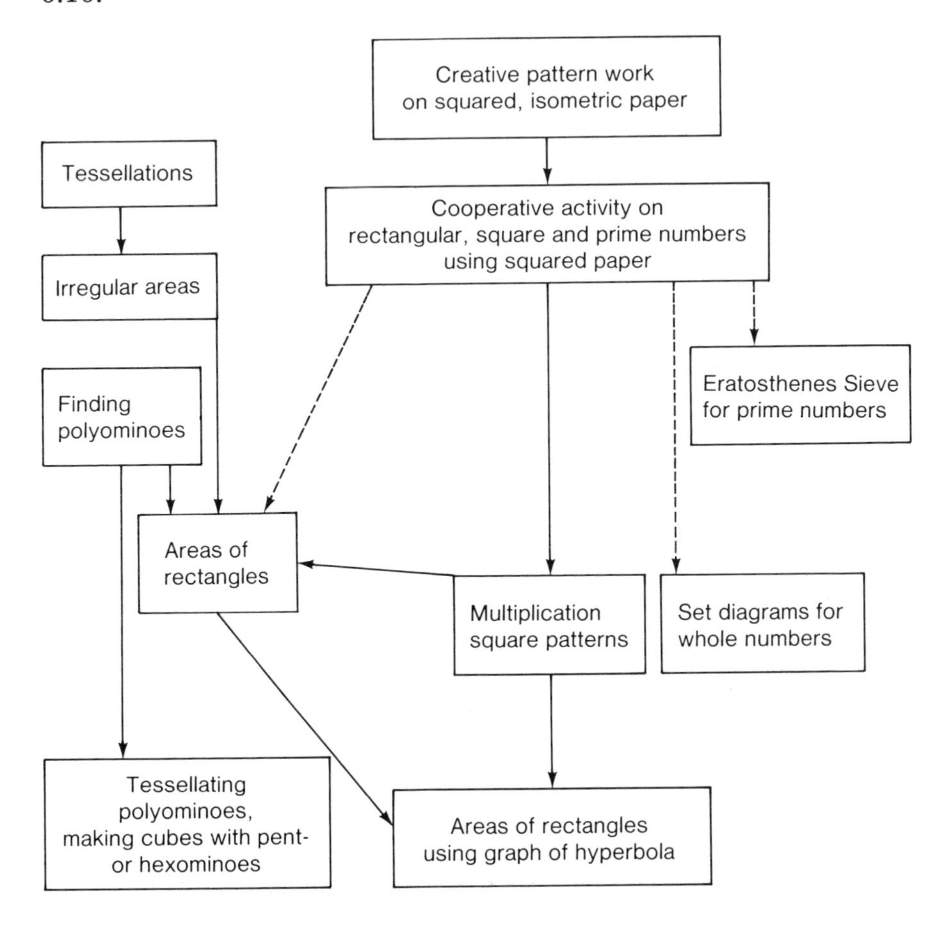

FIGURE 6.16

Improving a situation – heights of triangles

It may well happen that a mathematical situation is less effective than it might be in enabling the children to build and test their ideas. In this case we will want to strengthen the elements of doing, talking and recording that I introduced in Chapter 2. I shall take as an example the extract about the 'heights of triangles' from the same chapter, in which the teacher seemed to find herself forced into doing most of the mathematical thinking, rather than the children. What seems to be lacking is the *doing* element of my icon. However, mathematicians are extending the meaning of the word 'height' in an unusual way when they begin to apply it to triangles or other shapes. Questions like: 'How high is it? What is the height of . . .?' usually apply to heights above the ground, measured in a vertical direction. This is where the teacher might begin, by asking questions like 'Do you know your height? How would you measure it if you didn't know it?' and then continue with questions about the heights of trees, a building and finally perhaps a hill or mountain. The latter begins to make the discussion more abstract, and should highlight the idea of a base from which we measure. But what direction should we measure in? Does it have to be vertically from the ground? Why?

In the proposed new form of the activity, the class is organised into groups, each supplied with several cardboard triangles in different colours, or else these materials are supplied to the particular group working on the topic. One triangle is scalene, one obtuse-angled, and an isosceles or equilateral triangle may also be included. Sheets of plain paper and rulers are also needed. The teacher can begin with the following question:

> 'Hold your ruler on the paper and slide the (blue) triangle along it. What path does the top of the triangle trace out? Can you draw it?'

If the activity is with the whole class, an overhead projector will be very useful in comparing outcomes so that all pupils can see the various demonstrations. The blue triangle is the scalene one, of course.

> 'Now turn your triangle round and slide it on the other two sides. I want you to find – and mark on it – which way round the triangle is highest and which way round it is lowest.'

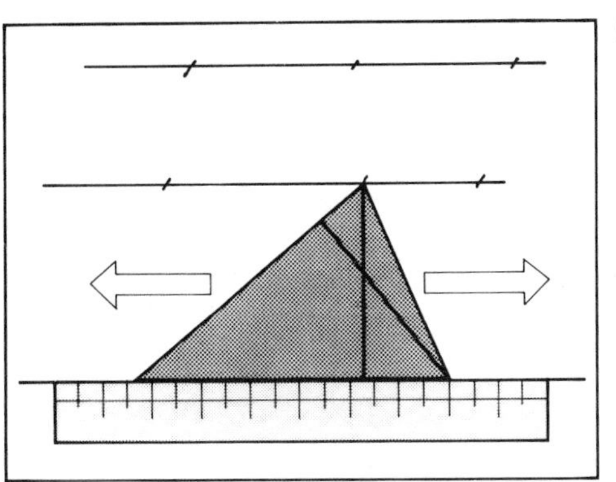

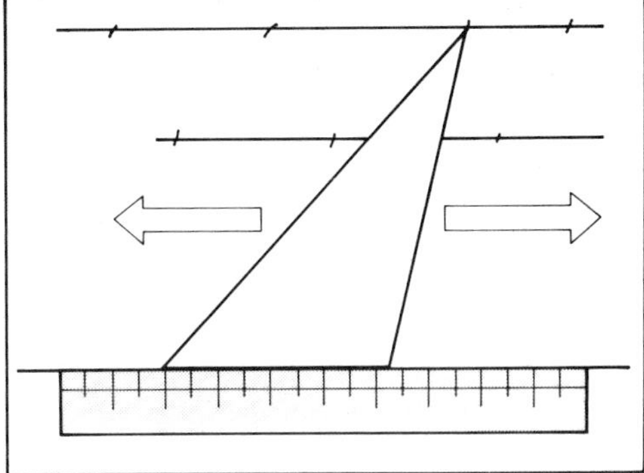

FIGURE 6.17

After more discussion, the 'heights' can be marked in on the triangle as in Figure 6.17. The investigation is repeated using the differently-coloured obtuse-angled triangle. As before, three parallel lines will be drawn, and so there are three 'heights', each corresponding to the distances between the parallels. But this time only two of them can actually be marked on the triangle. It is the distances between the parallels, called their 'gauges', which measure the heights of the triangle. This situation attempts to draw the idea of height away from a *particular line* associated with the triangle, which is properly called an 'altitude'. It also suggests that 'height' involves the notions of (*i*) choosing a suitable base to measure from and (*ii*) measuring in a direction perpendicular to this base.

Further questions might be asked about the other triangles included in the set of shapes – what can be said about a triangle when there are only two lines when it is slid along, and so two heights? When there is only one? 'Height' is being used here in the sense of a measurement, rather as the circumference of a circle is a measure of its perimeter. The situation can be extended as an

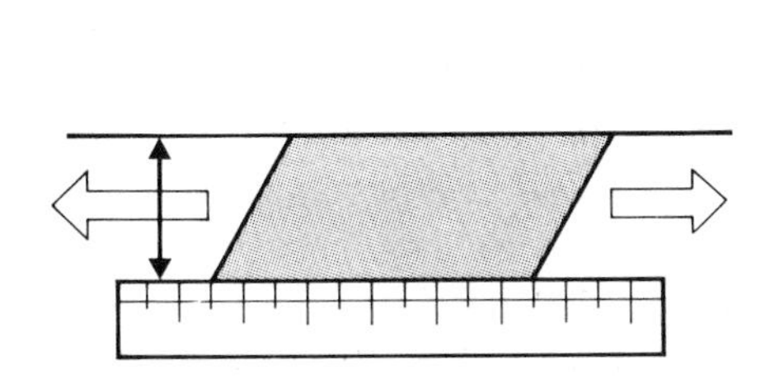

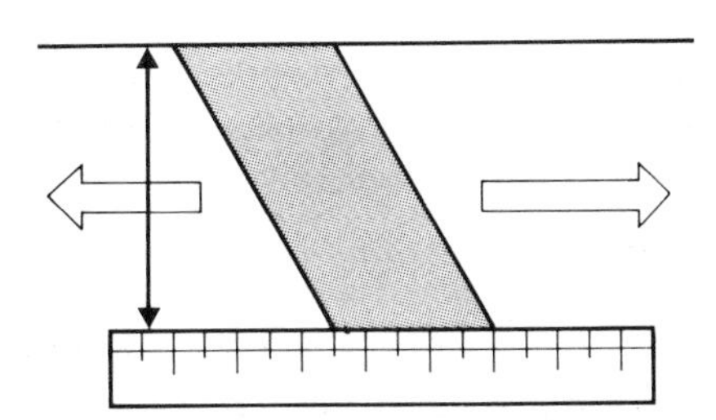

FIGURE 6.18

investigation using cardboard shapes, this time of parallelograms and possibly even quadrilaterals, as in Figure 6.18. Each one of the sides can be chosen as a base in turn, so that there are four possible heights in the case of the quadrilateral. I leave readers to find the other cases!

Investigation activities

Most of the ideas for mathematical situations discussed so far are on topics that fall within a typical scheme of work - number relationships, prime numbers, shapes, heights of triangles and so on. The Cockcroft Report also demands the use of investigations, which need not be about topics within the scheme proper. We have moved towards the open end of the dimension which I constructed at the beginning of this chapter, and are concerned with the two categories of aims entitled 'General strategies' and 'Personal qualities'. These are linked with what are termed 'process objectives' as distinct from 'content objectives', which relate to specific mathematical concepts or skills. Developing understanding of particular mathematical concepts, or giving practice in particular techniques, are not, it seems to me, foremost in investigative or exploratory activities - indeed, one might regard investigations as demanding applications of existing mathematical skills. General problem-solving skills, confidence, persistence and imagination seem to be at the heart of the matter. Among the former I would include pattern-finding, conjecturing or predicting, verifying or disproving, extending and generalising as important components. Among the personal qualities, besides the three already mentioned, coping with 'being stuck' and being able to accept and offer constructive criticism also seem important, particularly in group activities. Being stuck is a very common experience in problem solving, but much of our teaching seems to ignore it, or pretend that it does not exist!

Teachers can use various sources of ideas, either from the good books now available, or by following-up leads from the main topics. Mottershead (1978, 1985) has numerous suggestions, and the Lakatos group found SMP's *Pointers* (1984) particularly useful. Pupils may well find their own starting points from a topic such as the one on prime numbers described earlier. The calculator or micro may be helpful as a tool in some investigations, as I suggested in Chapter 5. Or it may offer a starting point microworld, such as 'Raybox', 'Bees', 'Strategy' and 'Colony', the four programs in the MEP's pack *Problems and Investigations Using Microcomputers as a Resource*. Sometimes a lead may be found in another area of the curriculum, such as art and craft or environmental studies. We tend to think of 'mathematics across the curriculum' in the sense of applying mathematical ideas to other areas, but the application could go the other way, in the sense that it generates a mathematical problem which becomes the source of a mathematical investigation. For example, art designs could start a study of symmetry (including perhaps some Logo programs), model-making an investigation of enlargements. Children's own hobbies or interests could form other possible starting points.

However, in the early stages teachers will probably present starting points selected by themselves to a small group or to individuals. It is at this stage that discrepancies between the teacher's and the pupils' perceptions of purposes, which I discussed at the beginning of the chapter, may well arise and perhaps prove troublesome. The children are likely to assume that the aim, as usual, is to learn some specific mathematics from you - which, as we have seen, is not the case. It is particularly important to gain and sustain their interest in the activity, and of course it would lose its purpose if they stopped doing the thinking! The approaches I have described in Part Two will form your means of achieving this. Interventions on such lines will convey a clearer perception of purposes to the children as they gain experience.

Objectives 15 to 24 in *Mathematics 5-16* give the detailed overview necessary for the teacher, but these objectives will only gradually become more apparent to the children. The teaching method consists of actually carrying out investigations, combined with hints and suggestions from the teacher at appropriate stages during the activities. Shuard (1986a) describes in more detail the varied approaches that have been tried in the past to teach problem-solving skills. These will repay study and perhaps staff discussion led by the mathematics coordinator. I will attempt to illustrate some of the points using a well-known investigation called 'The Tower of Hanoi'. Three discs, of decreasing size, are placed in order of size - with the smallest at the top - on the first of three pegs as in

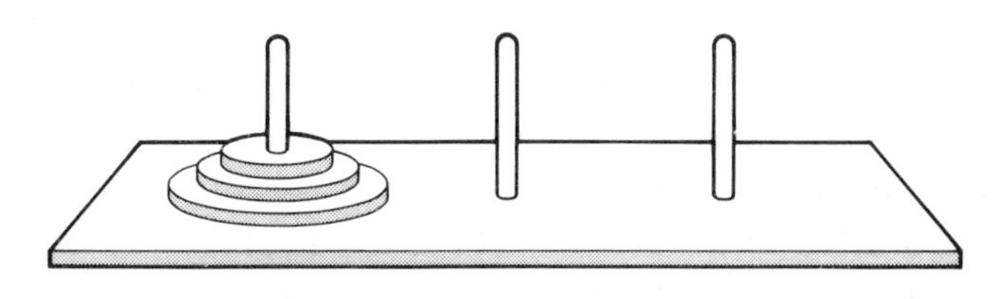

FIGURE 6.19

Figure 6.19. The discs are all to be transferred from the first to the third peg, following the rules that (*i*) only one disc may be moved at a time, (*ii*) no larger disc may be placed above a smaller one on any peg.

It is clear that the actual mathematics in this problem is not something that must be learnt as a topic in a scheme of work. Its purpose can only be discerned by relating it to the aforementioned objectives in *Mathematics 5–16*. Recently I was assessing a 'solution' to this particular problem from two children which consisted of lengthy lists of moves for every case up to seven discs on the pegs. At this point they gave up, finishing with the comment 'It has been a very hard problem to solve'. What had they displayed, of the skills and qualities listed? Persistence, certainly, and perhaps coping with being stuck on numerous occasions as they toiled with successive moves. Perhaps they worked well together but I was not able to judge this, as the assignment was a written one. Most notable was an absence of any search for or recognition of pattern, or any helpful form of recording.

How could a teacher intervene to help? In several ways, I think, through the approaches we have developed so far – not, it hardly needs saying, through extensive exposition, explaining 'This is how you do it', or taking over with 'Guess what I'm thinking'. Several of the things to be taught are habits that you hope the children will acquire from your own approach. 'Have you found an easy case?' is one of the simplest, and one to apply here. Closely related to it is the idea of investigating several instances, to try and 'get the feel' of the problem. It is important to get started on something, even if it seems very banal or trivial, and not remain frozen. This idea is also useful for getting oneself 'unstuck' on other occasions. By investigating the cases for two and three discs, which are not hard to solve by trial and error, children can build such basic experience. Now, rather than just finding a way of solving the problem, we need to find more systematic approaches and look for any patterns on which conjectures might be based. At this point the situation is very similar to that of the two children in 'Leapfrogs', in Chapter 3. Goldin, whose ideas I shall return to in Chapter 10, has shown that helpful forms of recording are extremely important for the two purposes I mentioned in discussing that extract, namely in finding patterns and keeping track of attempts. The sequences of actions are quickly forgotten, however successful, and so the next stage, which may require teacher intervention, is to develop some way of recording what we have done. These recordings can act as a permanent reminder, and also be compared with one another to

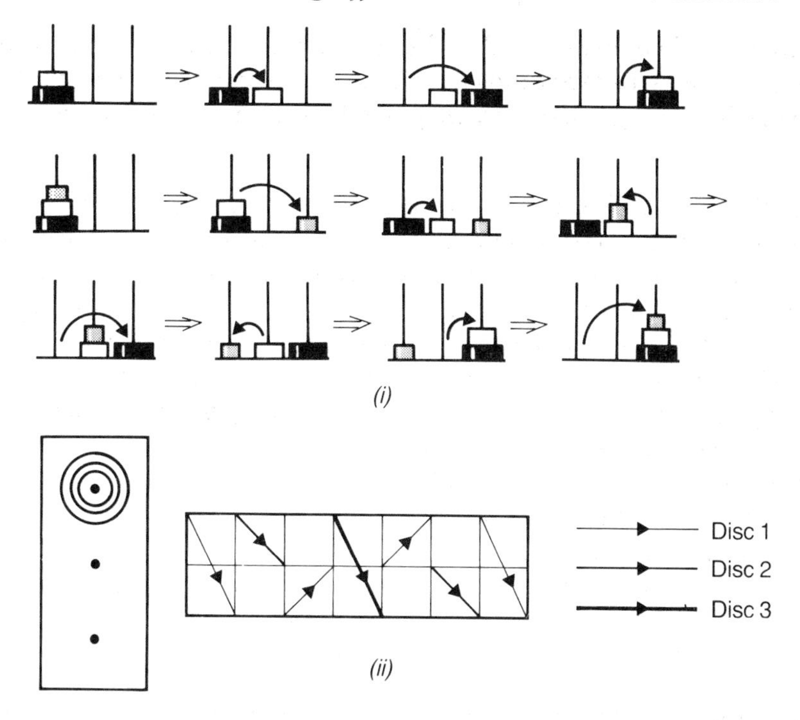

FIGURE 6.20

build an overall picture of our attempts on the problems. Thus with the Tower of Hanoi, a naive iconic form of recording might consist of drawing sequences of diagrams showing how the discs move on the pegs as in Figure 6.20(*i*). This is usually too close to the enactive situation to be very helpful. So in Figure 6.20(*ii*) I have developed a rather more abstract form, in which the pegs are represented across the diagram and the moves are shown by arrows, but stretched out so that they do not obscure one another. These arrows might be coloured so as to correspond to a particular disc. At once we see hints of symmetry about the moves, which closer study will reveal. The smallest disc moves four times, the next one moves two times and the largest just once. Aha!

How will the discs move when there are four of them? Can we use our diagram to help? For example, it seems likely that the largest will still move just once. When will that be, and how many times will the others move? We could persist with the present form of recording in Figure 6.20(*ii*) or invent a still more abstract notation, using symbols this time. But you can perhaps already sense the symmetry around the move of the largest disc – we have to transfer all the other discs to the middle peg, move the largest and then transfer the others by *the same sequence of actions as before* to the final peg.

I shall label the discs 1, 2, 3 in size order and the pegs A, B and C. The solutions in fewest moves for two discs and three discs then appear as in the table shown in Figure 6.21. Once again you can see the symmetry emerging in the patterns. Supposing I now want to use my solutions for two and three discs to help in the case of four. It looks as if the biggest, disc 4, will move just once, from peg A to peg C, and for that to happen I must have the others all properly placed on peg B. This is suggested by the first partially-completed line for 4 discs in Figure 6.21. I guess this will require 15 moves altogether, and this prediction could be based on two distinct ideas. First, I simply note that I will need seven moves on each side of the disc 4 move, to transfer the other three discs, first from A to B and then from B to C, and 7 + 1 + 7 = 15. Or I notice that in the previous solutions, the numbers of moves made by each disc have a pattern to them from largest to smallest disc – (1, 2), then (1, 2, 4) and so predicting for 4 discs, (1, 2, 4, 8) giving 15 altogether.

2 discs	Disc moved	1	2	1
	To which peg	B	C	C

NOTE THE PATTERNS

AND THE SYMMETRY!

3 discs	Disc moved	1	2	1	3	1	2	1
	To which peg	C	B	B	C	A	C	C

4 discs	Disc moved	–	–	–	–	–	–	–	4	–	–	–	–	–	–	–
	To which peg								C							

4 discs	Disc moved	–	–	–	–	–	–	–	4	1	2	1	3	1	2	1
	To which peg								C	C	A	A	C	B	C	C

FIGURE 6.21

In order to fill in the complete solution of moves from my partially completed line, I need to get the other discs *from* peg B *using* peg A. I can use my three disc solution from above, but have to remember that pegs A and B have interchanged their roles. I predict that the solution is in the second partially-completed line – very much as Eleri and Emma could predict using their pattern in 'Leapfrogs'. Can you now predict how the remaining half of the solution will be completed? You can test the full prediction by using some cardboard discs or coins. Notice once again the different role that the materials are playing – they are being used to test rather than to build ideas.

Investigations, then, should not consist of blundering about in a problem amassing particular cases, or hoping like Micawber that 'Something will turn up'. Teacher interventions need to be directed at ensuring that children are developing the general strategies described in *Mathematics 5–16*. Questions are likely to take very general forms such as 'Have you tried any simple cases?'. 'Show me what you've found out' might reveal that some kind of recording needs to be developed, so that further discussion can wait until this has been done. Tables are a very good resource in this respect, but not, of course, the only one. 'Are you using any plan? Have you seen any patterns?' are questions that might come at a later stage. Only when the children seem to have built some ideas from their plan of attack is it appropriate to look for predictions and ways of testing them. Some children may be able to extend or generalise the problem – for example Eleri and Emma go on to five and six 'frogs' in their game. Sometimes this brings some delightful and unexpected insights which stay in the memory and are worth sharing with other children, perhaps in the form of a classroom display. But I do not think one should get too anxious if this does not happen often! Unless children are really stuck and getting discouraged, let them work at whatever level their abilities allow, bearing in mind that no specific mathematical outcome is required. In any case, if children are badly stuck it may be better to change the problem. Coping with being stuck, and techniques for 'getting unstuck', have attracted attention recently – Mason *et al* (1982) offers a variety of ideas which might be of interest.

Thus, to summarise, we need to teach the idea of formulating some plan of attack on the problem, based on initial experience, from which a helpful form of recording what has been found can be developed. Children can develop their own forms in the first instance, but these might be modified by suitable interventions or later experience. From discussion and reflection based on their recordings, children *a* develop conjectures; *b* make predictions which are tested; *c* verify or disprove these predictions. Finally there is the possibility of *d* generalising the patterns or conjectures to other situations. There is no reason why the use of this language should not be introduced and developed among the children. The various stages of an investigation are suggested in Figure 6.22.

I conclude with a couple of investigations from *Pointers* which have worked for us, the first with infants and the second with

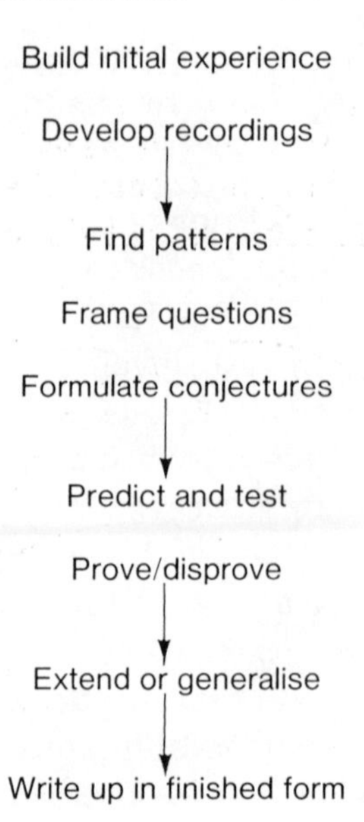

FIGURE 6.22

fourth-year juniors. You might like to try them yourself, and then think about your solution process in the light of what has been said, before experimenting with groups of children. Remember to make records and reflect, as described in Chapter 4!

Crossings

Figure 6.23 shows 2 parallel lines crossed by 2 parallel lines, then 2 lines crossed by 3 lines and 3 lines crossed by 4 lines. Can you continue the table which has been started?

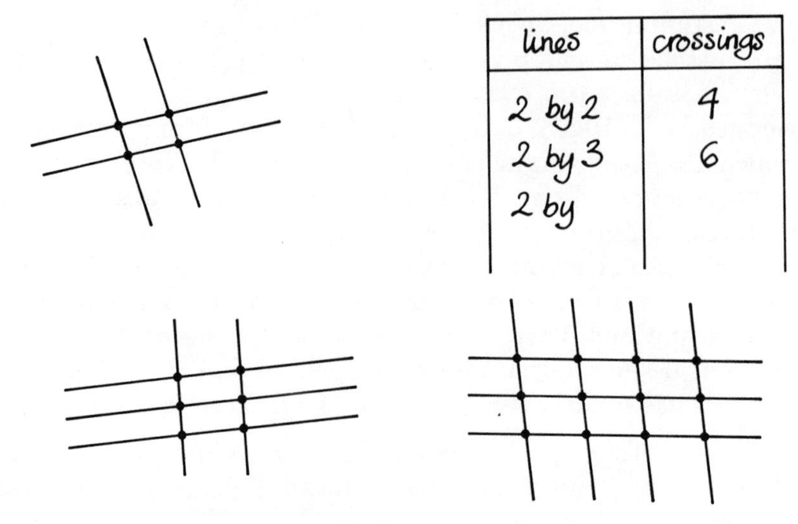

FIGURE 6.23

Adding ups

Insert the numbers 1 to 6 on the spots where the three circles in Figure 6.24 cross, so that the total of the numbers on each circle is the same. Can you find other sets of six numbers which will do this?

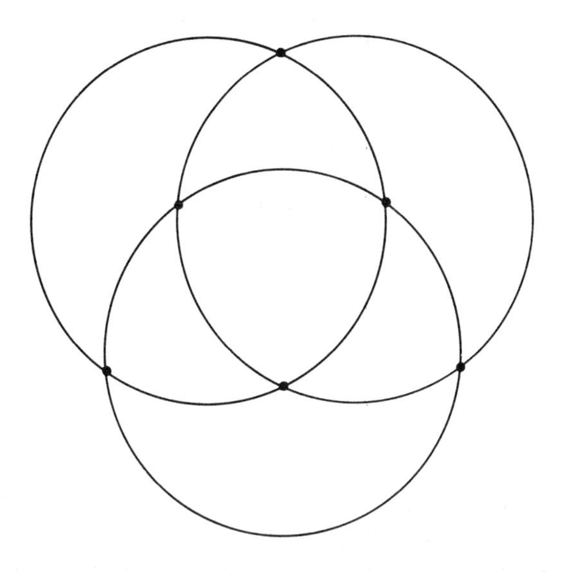

FIGURE 6.24

7 Groupwork management and organisation

The management task

An essential corollary of the developments described in Part Two is the setting up of an organisation to support continued groupwork in mathematics. I shall begin by distinguishing between groupwork and group organisation. A group organisation allows the teacher to set common tasks to groups of children, but they may, and evidently often do, work at these tasks individually. Such an organisation is a condition for groupwork, in which the children work cooperatively at the task(s) set. If we are to generate talk and discussion in mathematics, then both developments are required.

Boydell (1981) found in her studies of primary schools from 1970–77 that most of the teachers in the sample used a mixture of methods covering whole class work, group organisation and individual working. 'Unstreaming' was the main development, meaning that primary teachers everywhere were teaching the full range of ability. This presents formidable management problems, and nowhere more so than in mathematics. Thirty children following individual paths through a scheme are a severe challenge which outsiders seldom appreciate. Children are often seated in groups to ease the management difficulties. These groups are selected by ability or achievement and also, Boydell reports, by the ability of the children to 'get on well together'. But in the case of mathematics, joint tasks are seldom set and children mostly work individually, with the teacher relying heavily on written materials in her scheme.

I have a strong impression that infant teachers use a cooperative group approach far more than teachers at the junior level. Obviously they cannot rely on the use of written materials as the children are still learning to read, and so they make far more use of oral methods. It looks as if junior schools could profit from taking over some of their ideas, both with regard to organisation and to use of talk. At the junior level there are differences of opinion, with some teachers using groupwork extensively and others relying on individal approaches, although a group organisation may be used to manage the setting of the tasks. Moreover, these teachers may be in the same school! Besides these differences in outlook, schools differ widely in other important respects - resources, room sizes, open plan layout or traditional buildings, staff qualifications in mathematics and so on. Because of this it is clearly impossible to press some particular approach to groupwork on all schools, and so my main suggestion is that, following initial work on the lines set out in Part Two, staff decisions about forms of group organisation need to be made. On this basis, groupwork in mathematics can be developed.

It may be helpful to bear in mind some general points which can be discussed, in relation to the given constraints in a particular school.

1 *Are there members of staff, or infant teacher colleagues, whose expertise could be used to help and advise?* Where such advice is available, it could clearly help to get groupwork going, through staff development, even if the initial system needed later modifications. I shall offer some case studies later to indicate the possibilities.
2 *What are the problems over space, noise or resources? Could these be overcome by timetabling, team teaching, staff exchange, or other ideas?* Teachers can be inhibited from encouraging discussion because of the disturbance that might be caused to adjoining classes. Half a dozen groups may all be working at their tasks actively but creating quite a lot of noise. This might be overcome by having only one group at a time involved in a discussion activity.
3 *What changes in the attitudes or perceptions of the children might be needed?* Mercer (*op cit*) discusses the case of a class of seven and eight-year-olds who worked on a social studies project called *The Island*. One of the activities was a simulation game in which the children explored social relationships in organising themselves as castaways. The teacher's aims were all to do with general group skills, and the children worked enthusiastically and quite evidently practised these skills. But their perceptions of what they had learnt were quite narrow - 'It teaches you about what would really happen if you were on a desert island' or 'If you were stuck on an island and you wouldn't know what to do, so it gives you a lot of practice in that'! I discussed this difficulty in Chapter 6, in connection with investigations, and it

could clearly arise in this further context of group skills. Such changes in children's perceptions might take time to develop.

4 *What are the advantages of a joint staff approach to the planning of groupwork, rather than leaving the matter to individual staff initiatives?* The importance of this is obvious in relation to points 2 and 3 above. An individual, however talented, can feel quite uncomfortable working against the grain of a school approach. It would be far better to have a long term and generally agreed approach to groupwork, so that the children experience consistency rather than wide variability in teaching styles. It would give a longer period for their cooperative skills to develop, and more time for their own perceptions of aims to mature.

Any organisational scheme which supports the kind of groupwork we envisage in mathematics must include some vital features as follows:

a The teacher must be able to spend extended, largely uninterrupted, periods working with each group at some stage in a planned way. This is in addition to the brief interactions which occur on an *ad hoc* basis in any lesson, of course.

b The activities must be based on situations which engage the children for such periods, without the teacher needing to intervene in a major way. This is the property which I discussed at length in Chapter 5. At least some of these situations will involve discussion amongst the children without the teacher's presence.

Groupwork can be approached in a variety of ways. In one, the teacher carefully plans a number of mathematical activities which are all related to the current topic. Groups are formed according to some criterion such as ability or getting on well together. Each group works through the activities in the same sequence but with a staggered start. Less able groups take more time to cover the activity, while extensions are set for the more able. This plan allows the teacher to spend time with each group in turn. Later contacts are planned in a similar fashion.

Another approach, popular with junior teachers, uses a whole class 'starting point' lesson which sets the topic off into groupwork. The teacher then spends time teaching each group in turn, possibly starting them into the planned sequence of activities described in the first approach. In yet another approach, the class is allocated to groups and only one of these works on mathematics at any given stage. The teacher can therefore choose to work with this group, setting tasks for the others which allow them to get on without her intervening. This pattern is popular in infant classes. The tasks may be related to different levels of ability, particularly at the junior stage, rather than planned as a common activity. If transferred to the junior stage, the approach would involve a change in the common timetable pattern in which all the class do mathematics at a particular time of day, usually the first lesson. Teachers who use a groupwork approach, particularly infant teachers, often express surprise when I suggest to them that there should be any particular problem over its organisation. Yet, as I said in Chapter 1, the commonest question I get over discussion is 'What were the other children doing while you were discussing with this group?' (or words to that effect). I think that there is a problem, which seems to be partly to do with teacher attitude and partly to do with meeting the two criteria for the mathematical activities which I set out above. Evidently it would best be overcome by a joint staff approach, as I suggested. Let us see how some particular teachers approach the matter of organisation.

General organisation in an infant classroom

This teacher has a class of 27 top infants (six to seven year olds). They are organised into four groups based on their mathematical ability, and identified by colour. The Yellow group is above average, the Red and Blue groups are average, while the Green group is below average and experiences reading difficulties. The children work from a task card, when they are not involved in a directed teaching activity within their group. These cards contain tasks which the children are expected to accomplish through the sessions, for example:

Phonic work
Language workbook or cards
Handwriting skills
Story or news writing
Mathematical activity – such as

apparatus to explore
shape work, patterns
practical measuring
work to do with time
constructional apparatus

The classroom is organised to facilitate the groupwork in language and in mathematics. The teacher plans her timetable so that she teaches each group directly at some stage in both of these aspects. In any one session she works with two groups, one in mathematics and the other in language. The other two groups also work in this way, but in activities which are 'less demanding of my valuable time'. Thus every child gets some specific period of time with direct teaching from the teacher. She also allows time for individual attention, particularly in the Green (lower ability) group which is given an individual programme of work.

Her school has specific space arrangements, like all schools. Her classroom has the advantage of sharing an adjoining room with another class. Here it has been possible to set up areas for sand/water play, creative activities and a science discovery area. There is also a computer corner where her children are timetabled for use of the single BBC computer. A 'home corner' has been created, whose displays are changed periodically, and this is used to foster language and social skills. Perhaps it could also be used to foster mathematics skills?

Her own classroom, very typically, contains a 'mathematics area' in which all the apparatus, books and resources the children require for their activities are stored. Use is made, again very typically, of displays of children's work, charts, models and so on. There is a large carpeted area where children work with constructional apparatus (as in the *Clixi* account in Chapter 5) or play floor games based on logic materials which I have also described.

Occasionally situations develop which require that the children work in pairs, discussing the problem involved. This works well in practical investigations such as weighing and measuring activities, sand/water play or in the shopping corner. One partner is involved in the operations while the other records findings as their work progresses. 'Real discussion', says this teacher, 'wherever it appears is provoked by experience. It may arise spontaneously or be provoked by the teacher's interventions. But the situation supplies the starting point and the discussion that ensues should widen the children's horizons and open up further avenues to be explored'.

I challenged another infant teacher with that familiar question 'What are the other children doing?' Again there was expertise in the handling of groupwork and its organisation. She also worked with four ability groups, whose daily tasks were displayed on charts or a 'wheel'. At the start of the day she explained which groups she would be working with, organising her contacts in very similar fashion to the teacher in my first account. The rest of the day followed the pattern shown in Figure 7.1 – this applied from Monday to Thursday. Friday was a kind of 'tidying up of loose ends' day – the morning devoted to whole class work on science, mathematics or language, the afternoon to play or reading along with 'sorting out the chaos of the week'. Anyone who has experienced the aftermath of art, craft and practical activities by lively children will know what this involves! 'I am a timetable sort of person', she says, 'and work better when I'm organised'. Of discussion she continues 'I feel that it should be an integral part of the task, rather like practical work. The teacher *who is aware of the need for and desirability of discussion* will not need prompts'.

9.00 – 9.10	Greetings and registration
9.10 – 9.30	Assembly (computer programs loaded from tape in this time)
9.30 – 9.40	Discussion of day's tasks (including which groups teacher will work with)
9.40 – 11.45	Tasks 1–3, with a break in the middle
11.45 – 12.00	Discussion of tasks
Lunch	
1.20 – 1.30	Class lesson (eg spelling, number bonds etc)
1.30 – 2.30	Tasks 4 and 5
2.30 – 2.45	Break
2.45 – 3.30	Two activities from Singing, Story, Music, RE, Dancing

FIGURE 7.1

This is an account that will be familiar to many infant teachers, I am sure. Here is a similar one from a junior school teacher.

General organisation in a junior classroom

This teacher's class consists of 25 fourth years plus 3 third years, 12 of whom are girls and 16 boys. In ability it ranges from children well above average to some who attend a remedial reading class and several with other learning or behavioural problems. The children come from working-class backgrounds, living mostly on a post-war council estate. Problems such as one-parent families, unemployment, even poverty, are quite common, yet the children are materially well-catered-for. Every home has a television set, most have videorecorders, 12 homes have a computer and most of the children possess their own calculator. The children's main deprivation is the lack of communication at home, and many have a very limited vocabulary, she reports.

Her classroom is in a typical 1930s building, one of a row along a corridor. One wall is windowed, the facing wall has narrow, high windows. The furnishing is an odd mixture of modern tables and storage trolleys along with venerable cupboards and bookcases that have weathered many years of service. The room is set out as in Figure 7.2, like many other junior classrooms today.

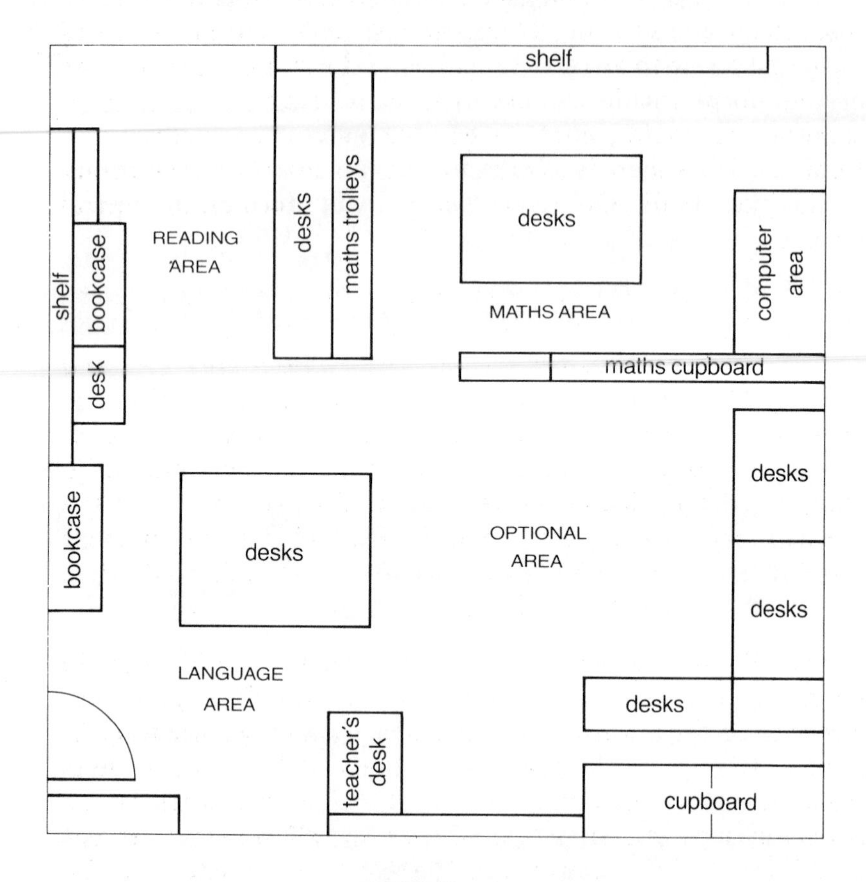

FIGURE 7.2

This teacher's grouping methods are of interest. She writes at some length as follows:

> Initially the composition of my groups, usually four in number, is decided on mathematical ability – the outcome usually coincides well with the use of the alternative criterion of language ability. Their mathematical ability is assessed using records and other information provided by their previous class teacher, combined with the profile provided by a standardised test, the *Richmond Profile of Mathematical*

> *Skills and Concepts.* But I also use compatibility in forming the groups. Their sociological structure does not always follow that in play or for physical education etc., although that knowledge is very useful to me. I try to make sure that there is a balance of 'theoretical and practical' expertise in my 'teams'. As they become established any differences become lessened and more often the children develop an unspoken but mutual respect for one another. This does not mean that all members wish to work with all others. Usually workgroups of two, three or four form within my organisational 'teams'. These meet together to confer, confirm or seek each other's ideas and findings. Thus there is a constant moving apart and reforming of the main body. Only very occasionally does an individual choose to work on her or his own.
>
> I have been with the children for two years, having followed them up from the third year, and so we know one another quite well. My system uses the various areas (see Figure 7.2), and also whole class lessons. We work in these areas at different levels in small groups – I plan at least four different levels of groupwork for each subject. My approach means I only need small sets of particular books or apparatus, so we can stretch our capitation money much more effectively, for example by providing extension material for the most able and some special equipment for the least able.
>
> I generally introduce new work in mathematics on Monday, in language on Tuesday. If the mathematics topic is ongoing into a second or third week, I may introduce a new science topic in my fourth (Optional) area. I have found it best, for obvious reasons, not to introduce new work in several areas of the curriculum on the same day. Tasks for all subjects are set weekly and task sheets for the week are posted in the appropriate area. Friday morning is a review session, in which I can gauge the work for the following week.

This teacher goes on to describe the importance of seating arrangements which encourage collaboration, of organising all the tasks and materials clearly in the different areas, and training children to help in keeping everything where it can be found and to respect others and their work. This social training she regards as vitally important. She uses the whole class approach mentioned above in starting a new topic. This draws on children's knowledge and interests, for example 'tessellations' might evoke father laying a patio, patchwork counterpane on a bed, packages in a box and so on. After such a general introduction, the class breaks into its groups, where she meets with them for further development at their level. This teaching takes place four times in rotation on the introductory day.

> What are the other children doing? It is easy to see why I only introduce one new topic at a time. Suppose on Monday it is mathematics, then I plan the other group activities to be quite self-supporting. Language could be a creative writing period, reading might be silent reading of class library; recording last week's class work on tape or a comprehension exercise. The optional area tasks might be personal topic work, research, handwriting etc. This allows me to concentrate on the new maths with each group in rotation – a typical Monday timetable might be like this:
>
> 9.40 – 9.50 Introductory whole class discussion to new topic
> 10.00 – 10.40 Group session 1
> 10.55 – 11.40 Group session 2
> 11.40 – 12.00 Welsh
>
> Lunch
> 1.00 – 1.40 Group session 3
> 1.40 – 2.15 Group session 4
> 2.25 – 3.00 Physical education
> 3.00 – 3.30 Whole class lesson (poetry/story/singing etc)
>
> In the course of a week, then, each group has four forty-minute sessions and Friday morning on the mathematics topic, to complete the task set and move into appropriate extension work. It is important that this extension work is planned and posted, otherwise children may feel that they are 'being punished for finishing the task early'. They can very soon learn to pace themselves to take just the four group sessions! Towards the end of the week the groups could be working at widely differing levels. The top group could be making quite complicated applications, the middle groups achieving success and confidence in using the new concept. The bottom group should have had ample encouragement from me and be confident in their use of apparatus and presentation of work. This could extend into their option time if necessary.
>
> I use the same general plan to introduce new topics in all curriculum areas, not just mathematics. I arrived at it after some experimentation – I do not believe I could teach the

whole of the class as a class so as to extend the most able, fulfil the potential of the large 'average core' and also cater for the needs of the less able, all at the same time. My approach allows me to concentrate my teaching on each group in turn in whichever area of the curriculum I introduce new work. I spend much less time in contact with the other groups. Because I encourage my children to 'do and discuss' I find that many of the minor problems never reach me.

The manner in which this teacher's planning develops is shown in Figure 7.3 – the Monday timetable. Her contact time is shaded.

Whole class introduction (maths topic)			
Group 1	Group 2	Group 3	Group 4
Maths	Language	Reading	Optional
Language	Maths	Optional	Reading
Whole class			
Optional	Reading	Maths	Language
Reading	Optional	Language	Maths
Whole class			
Whole class			

FIGURE 7.3

The 'staggered sequence' approach

This approach is described in some detail as part of the Open University course *Developing Mathematical Thinking* (in topic 3). It assumes that all of the children in the class will be engaged on mathematics at the same time, unlike the system adopted by the teacher in the previous case study. The class is organised into groups, mainly by ability, each group cooperating on a task planned by the teacher. The constraints are, of course, exactly the same as those described in the previous account, except that they relate to mathematics rather than to four separate curriculum areas. If the teacher is to concentrate her attention on a particular group for some length of time, the other groups must be relatively self-supporting. This can only be achieved, as before, by careful planning.

Each group will work at its own pace, and so extension material has to be planned, and more time possibly allowed for the less able groups to complete the essential activities in a topic. These activities are planned as a sequence followed by all groups, and extending over a long period such as a half-term or even a term. The groups are started into this sequence in succession over several lessons as suggested in Figure 7.4 – four if there are four groups, for example. This enables the teacher to spend time concentrating on introducing the new work with each group at the outset.

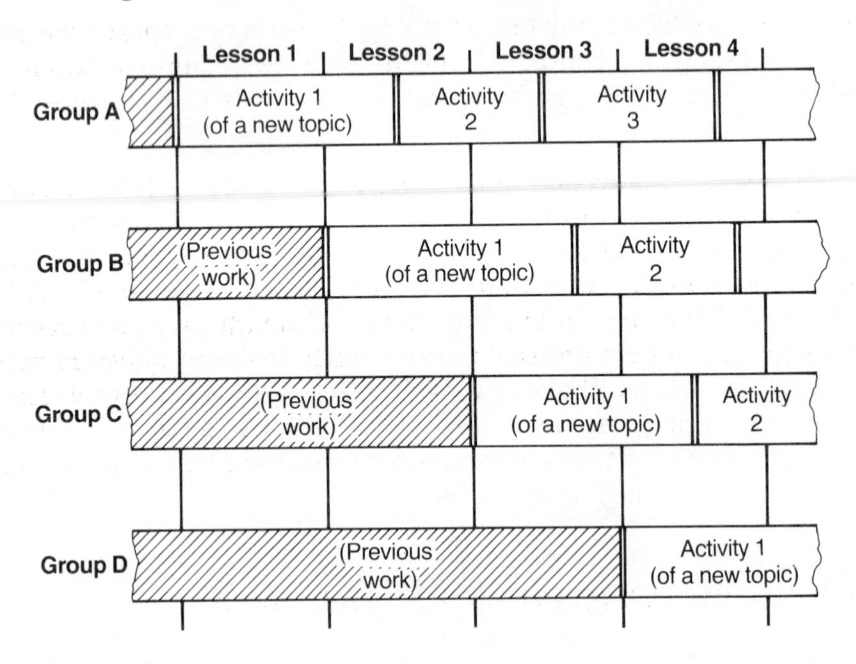

FIGURE 7.4

In subsequent lessons, the teacher's contact time is similarly planned with each group in turn. For this purpose the OU course offers a planning idea which may be helpful in other approaches, the 'teacher involvement strip'. This is simply a diagram (Figure 7.5) showing the sequence of activities planned for a group over several lessons. On it the teacher shades in the periods of time when concentrated involvement will be needed, because of the stage the children have reached. The result then has to be compared with the involvement strips for the other groups, to avoid over-commitment, and so some re-planning or extension work may be needed. An

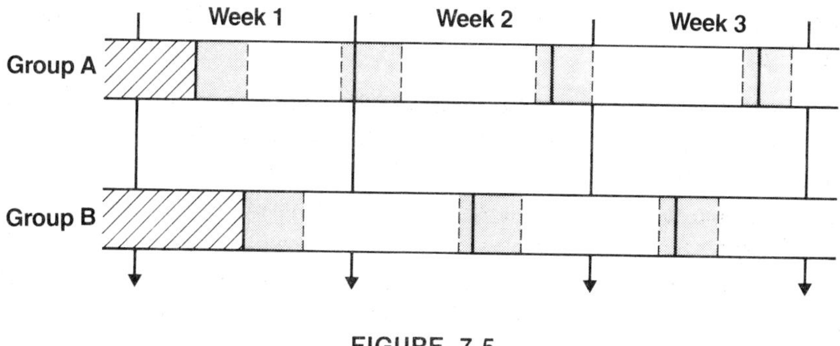

FIGURE 7.5

alternative which many teachers already use is to prepare a weekly timetable of group contacts from day to day. Accounts by teachers about the approach can be found in the course reader, edited by Floyd (1981).

Like the previous account, this approach obviously requires careful attention to preparation and planning. This is perhaps the main message that emerges from all these suggestions. But can this really be more difficult than coping with the demands of thirty children following individual paths through a mathematics scheme? Are we not entitled to ask exactly the same question with which I am so often challenged, 'What are the other children doing?' Is it not simpler instead to think of our class as a relatively small number of cooperating groups, and plan both the activities and our contact time accordingly?

Planning for groupwork as a staff

I have already indicated the possible advantages of planning as a staff to develop groupwork in mathematics. It is almost certain that language policy and the curriculum as a whole will be drawn into these discussions. I shall turn to these matters in Part Four. Meanwhile, in this session, I shall outline how one particular school tackled the challenge. Their ideas have contributed much to the suggestions put forward in Part Two for 'getting started'. The role of the headteacher was crucial in these developments. She held the view that continuity of approaches and organisations was essential throughout the school, but that it was vital to involve the teachers in the curriculum planning. Various teachers were given specific responsibility for different aspects of this process. For example the language coordinator set up ways in which objectives could be partially realised in mathematics, another member of staff monitored the effectiveness of the scheme of work and the use of discussion as an aid to understanding, while a third investigated the use of audio visual resources as an aid to discussion. Listening skills were felt to be particularly important, and all children needed to be familiar with the use of a cassette tape recorder.

The staff programme developed very much on the lines we have set out in Part Two. As confidence grew, teachers became aware of strengths and limitations, but also began to appreciate the effectiveness of their sessions. A variety of points were noted, to do with assessing children's use of language, their strengths and weaknesses, stage of mathematical development, and understanding of concepts. Effective mathematical situations that encouraged discussion were identified, but in addition teachers took a closer look at their own role. A list of self-critical questions was drawn up, including the following:

- How well did I plan for the session? How might it have been better?
- Did everything go as I expected – were there any surprises?
- Did I allow the children to think for themselves?
- Did I value all their opinions?
- Did the children understand the purpose of the activity?
- Was I clear about my aims and objectives?

You will recognise many familiar points here, that have raised themselves in previous chapters. This staff developed the method of sharing ideas that I have outlined in Chapter 6. They carried it further by making extensive use of cassette recordings, which were used as a basis of staffroom discussions about what had taken place. The recordings were also played back to the children so that they could comment on their own performance. This interesting application may be worth following up as a possible means of modifying and improving group interaction. From this exchange of ideas the staff began to build their own mathematical resource area, in which they filed examples of good activities under the area of mathematics covered. Their cards contained the following information:

- a photograph of the session
- proposed weekly notes for the session –
  How I planned the lesson, its organisation, the methods etc.
- a quick evaluation sheet
- full dialogue with comments and follow-up activities

The head comments 'It must be remembered that these cards have been produced to suit the needs of our school and match our form of organisation. They are written by our own staff and are the best way we can record ideas for future reference. Your cards may look quite different but should reflect the needs of your school.'

In rewriting the school's curriculum statement to incorporate 'mathematical discussion' the following framework was developed. First, various aspects were identified, as in Figure 7.6.

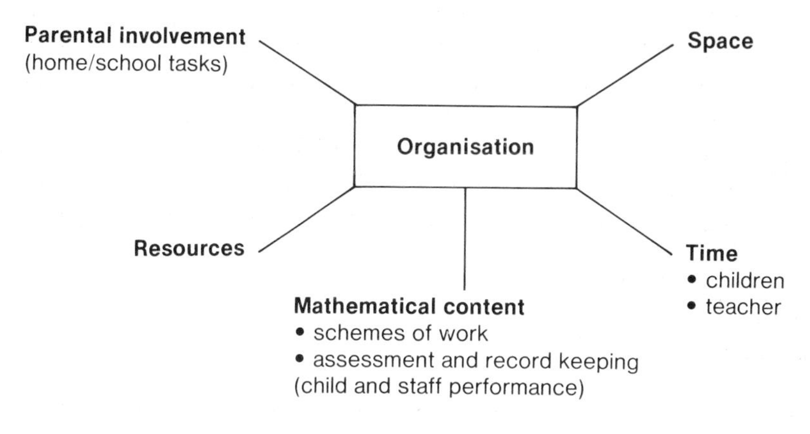

FIGURE 7.6

Each of these aspects was then developed in more detail. Under 'space', for example, the following suggestions were noted:

The organisation of the classroom should support the stated aims and objectives of the curriculum and provide opportunity for the child to develop

- contact and socialisation with other children
- language and communication
- self-expression
- learning how to learn
- learning in small groups
- learning in large groups
- problem-solving skills
- real life experiences
- cognitive and affective development

A typical mathematics area should offer opportunities for the children to work individually or as a group. In the plan shown in Figure 7.7 there is enough space for six children to sit facing one another round a group of tables and also a quieter area for three to four children to work.

Under the aspect of 'time', the children's and teacher's use of time are distinguished. The children's time is directed by the teacher or

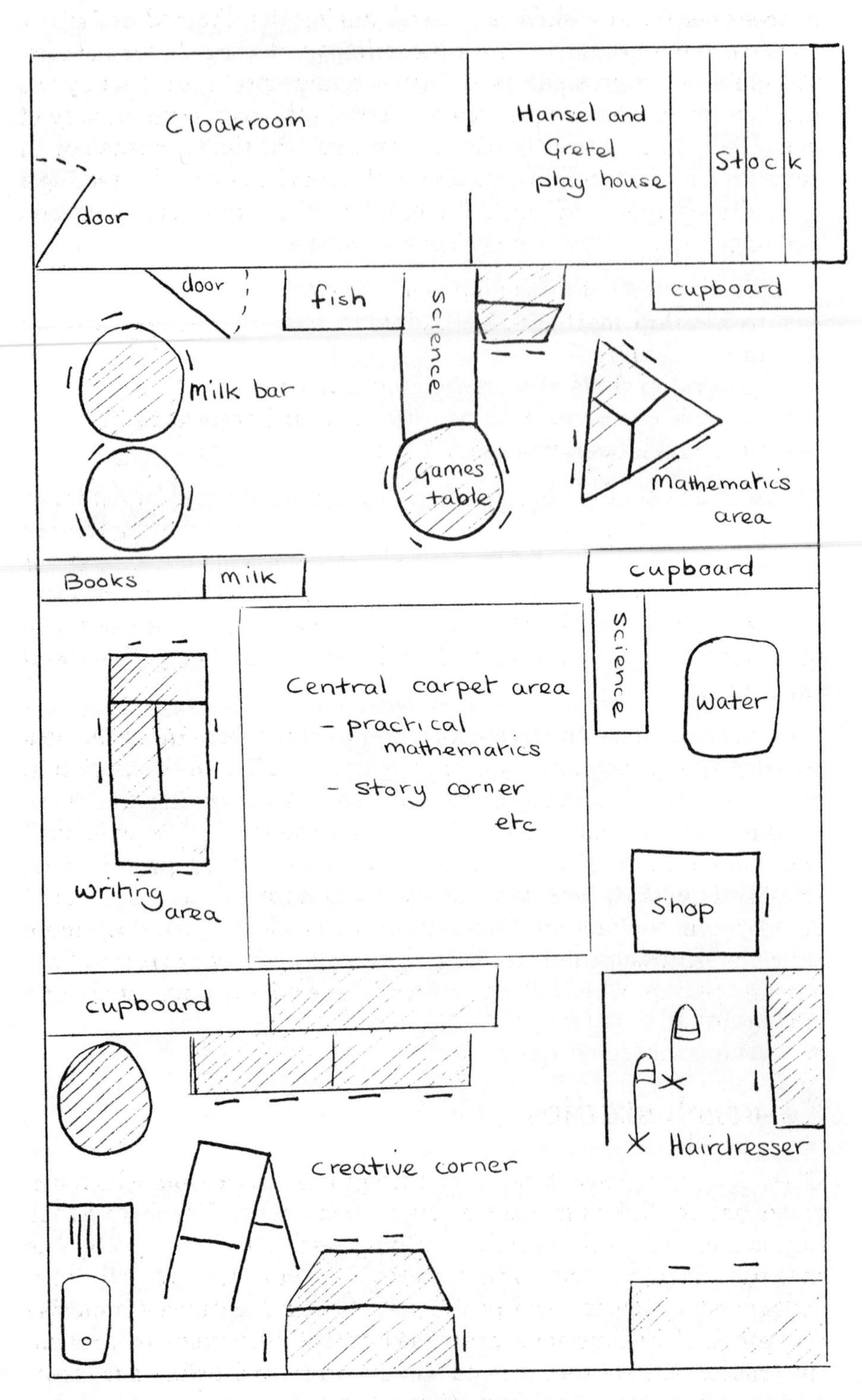
Cloakroom
door
Hansel and Gretel play house
Stock
door
fish
science
cupboard
milk bar
Games table
Mathematics area
Books
milk
cupboard
science
Central carpet area
- practical mathematics
- story corner
etc
water
writing area
Shop
cupboard
Hairdresser
creative corner

FIGURE 7.7

is 'free choice'. New skills are learnt during the directed activities, creative and exploratory work is carried out during the free choice. The children are grouped by ability level and taught together by the teacher during the directed time – 'This is the only sensible way of using her valuable time to best advantage'. Particular needs are, of course, identified and given individual attention, partly by setting a special programme of work. Several types of group are identified and organised for differing purposes, such as:

- for directed skill teaching (same ability);
- to develop mathematical thinking and discussion (same or mixed ability);
- to develop skills of working cooperatively;
- to share resources such as constructional apparatus;
- to assign a task, worked individually.

The teacher's time in 'directed teaching in mathematics' must focus on the use of discussion to develop meaning and understanding of concepts, before skills are introduced. Time must also be allowed for an assessment of children's progress. 'Confidence and success are the two crucial qualities that affect discussion . . . it is necessary to plan activities that are directly related to children's first-hand experiences.'

It seems clear from this account that the school has embarked on a constructive programme around the use of talk and discussion in mathematics. This programme focusses on improving the curriculum for the school's own children, uses the ideas of its own staff and takes account of the advantages and constraints specific to the school. The staff are also likely to benefit in terms of self-development – skills of conducting discussion, planning more effective situations and so on. So the programme offers most of the advantages described in the section on in-service training at the end of Chapter 4.

Research studies

There is a large body of research into group interaction, which has given rise to differing theories about behaviour. 'Psychodynamic theory', linked with Freudian views, sees the 'basic group' as sharing certain tacit assumptions or motives, to do with dependency, security and protection. Its explanations emphasise the effect of unconscious processes on the behaviour of groups. Interaction theory focusses on the interpersonal behaviour itself, using various forms of observation schedules to categorise it. Bales has been one of the most prominent workers to follow this approach. Groups appear in all aspects of life, not least in education, from the staff or departmental meeting to the meetings of governors or education committees. But we are concerned with a particular aspect, *learning in groups*.

Unfortunately, much of the research in this area is based on studies of adults or students in higher or further education. For example Jaques (1984), in a thorough overview, concentrates entirely on the latter. Only in recent years have studies of schoolchildren emerged, notably by Barnes and Todd (1977) and Tough (1977). Button (1981) developed a full programme of *Group Tutoring for the Form Tutor*, designed to encourage cooperative development of learning skills. The programme is aimed at secondary schoolchildren, as the title indicates. It was based on widespread action research studies by Button and teams of teachers and youth workers, over a period of several years. Tann (*op cit*) carried out a study using twenty-four groups of four or five children drawn from top juniors and first year secondary pupils in two schools. Each group worked on the same series of four tasks, all open-ended in nature. Two were factual, requiring reporting and reasoning, two involved fantasy and required interpreting and imagining skills. Other variables such as sex, achievement, and free or directed allocation to groups, were investigated. The resulting tape recordings were analysed using a category system. This had two types of category, one concerned with reporting, imagining, reasoning, evaluating and directing, the other with interactional relationships such as initiating, challenging and accepting. These categories were used to chart the problem-solving process in each group, identify its verbal strategies, the ways in which the participants worked together, and the nature of each member's role.

Tann's findings are of interest, particularly as they can be related to the age group with which we are concerned. She found two strategies to be of crucial importance: reasoning and challenging. Where reasoning was absent or implicit, discussions were not successful. Challenging was a strategy used to seek reasons or justification (Can you see its possible importance in our own special concern?). It was an aspect of *questioning*, which was not effectively employed in many of Tann's groups. This led to fewer alternatives being generated and tested and so a less successful outcome. In this connection Tann was careful to define success in terms of the group, not in terms of some predetermined 'solution' – 'A discussion was considered a success if the group appeared to solve the problem as they defined it, in a way that was felt to be satisfactory to that group'. Besides these strategies, listening and managing disputes appeared to be of particular importance. Some children found it hard to listen to others, and some groups talked at each other, generating repetition and competition about which idea belonged to whom. Personality and status were at stake, rather than concern with the task in hand - shades of the adult world! Some important differences between groups of boys and girls were noted. Boys seemed more willing to take risks, and treated ridicule as a joke. Girls were more consensus oriented, challenged less and left difficult issues unresolved. Slower boys joined in and were helped, while silent members were ignored. Slower girls tended to remain silent, despite appeals and support from the other girls in the groups. The leadership role tended to be less well-defined among the boys - with the girls a clear leader accepted as the 'brightest' by the others tended to emerge. Tann found the mixed sex groups full of problems, which became sharper with the older pupils. In fact, in the secondary school, she was only able to persuade one mixed sex group to take part in her experiment!

Jones, as part of his *Groby Oracy Project* (1985-6), carried out studies in both primary and secondary schools. In one of these he studied attitudes to talk as a learning activity among 52 second and third year pupils in a secondary school. The findings are somewhat discouraging in some ways, hopeful in others. Whole class discussion enjoyed only a small advantage over teacher explaining, while small group discussion, though mildly enjoyed, was rated bottom as a learning activity! When Jones probed more deeply, he found the pupils were concerned about control in such situations - they cited others 'messing about', or 'doing all the talking'. There were also complaints about the discussion drifting away from the task in hand. Despite all this, there was a strong undercurrent of appreciation of the values in such discussions, to do with participation, sharing of viewpoints and so on. The narrow view that many pupils have of learning has been remarked on in previous researches, eg Nash (1973, 1976). I have already referred to the need to attend to this potential clash between teachers' and pupils' attitudes and perceptions. But Jones' work reveals at least the possibility for positive change, if we go about it in the right way.

Research studies can throw little light on the organisational aspect, the main concern of this chapter. Much of the available expertise seems to have been acquired by teachers 'on the job', and I can find little published material or terminology, although Dean (1983) has written on the organisation of learning at the primary stage. Mathematics educators have arrived at this situation by a

8 Approaches to whole class discussion

Is whole class discussion possible?

Sceptics amongst my readers may doubt whether discussion, in the sense I described in Chapter 2, can be obtained in working with a class of perhaps thirty children. Certainly one cannot obtain full interaction amongst such a large number of potential contributors; only a relative few can actually speak to the class as a whole. The way forward, which can work very effectively, is to organise the talk at two levels. The class is formed into groups, which will discuss mainly on their own. The teacher interacts with these groups by circulating – this is the first level. He or she controls a second level of interaction between groups, by calling on spokespersons to report, and drawing in other children appropriately. Obviously this calls for skilful control, in a kind of 'chairperson's role', although that analogy is not entirely apt, as I have suggested already in Part Two. The groups will normally need to be working within the same mathematical situation, or else they will have no basis on which to discuss! Such a whole class situation will need choosing and planning with extra care, because it must be effective across the range of ability. Next the working groups will need to be organised as usual, with the criteria which I have already outlined in mind. Seating will need more attention, for reasons which I will go into presently. Last, a lesson pattern must be devised, to allow these two levels of operation. I shall describe two such lesson patterns in the following sections, and exemplify them with actual accounts.

Whole class activities will usually only occupy part of a given lesson period, and then lead into follow-up work in small groups. Teachers will clearly need to bring to bear all of the interaction skills they have developed through working with these small groups. At all costs, we must try to avoid regressing, in whole class work, to the three-term sequence of question and answer, or any of the other forms that, as we saw in Part One, may bedevil and hamper discussion.

The starting point pattern

This widely-used pattern is familiar to many teachers and will need little description. It begins with a question or task being set by the teacher to the whole class, then a short session devoted to drawing out of children's ideas. It is normally used to start off work on a new topic, which continues on an individual, or sometimes small group, basis. An example of a possible lesson is the activity about 'Square and rectangular numbers' which I described in Chapter 6. Discussion continues within the small groups, and the teacher circulates round these, perhaps using the organisational systems which were described in Chapter 7. Teacher-pupil interaction takes place during these sessions.

The starting point lesson is straightforward to conduct, but the interaction will be improved if the techniques which have been developed in Part Two are employed in the whole class work as well, rather than the usual question and answer patterns. The follow-up activities may extend over several days. Groups or individuals will obviously take varying lengths of time and reach variable standards, which the teacher clearly anticipates and plans for, as was described in Chapter 7. They may present their work, in a neat

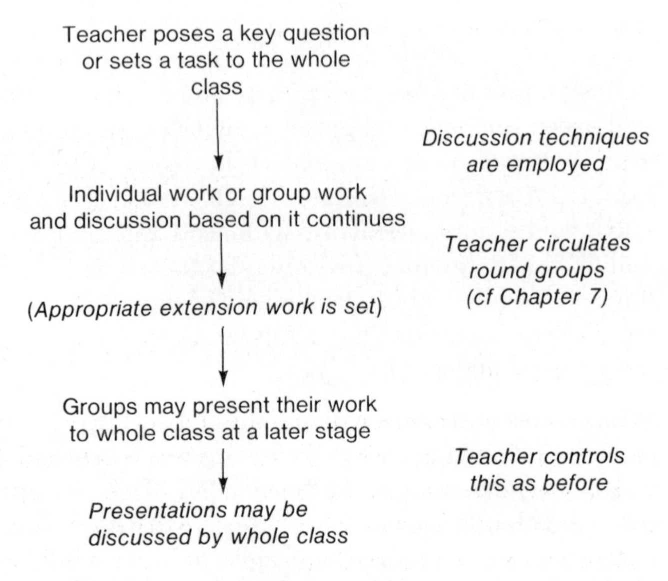

FIGURE 8.1 Starting point pattern

and finished form, at the end of the topic. It may also be helpful to provide an audience at times (following a suggestion made by Wiles as reported in Chapter 9) so that spokespersons can describe their ideas orally as well as in written form. This makes a good conclusion and a way of sharing, in limited form, what has been achieved. This type of motivation is well-known to primary teachers, but is not used in mathematics, perhaps, as much as in other areas of the curriculum. The starting point pattern can lead into very diverse activities and is an effective way of handling a wide range of ability, whilst retaining an element of sharing ideas. The general flow is shown in Figure 8.1.

Timetables – a starting point lesson sequence

The teacher in this extract uses the groupwork organisation described in Chapter 7. She teaches each group mathematics in turn, while the other groups are engaged on different, self-supporting activities. Her main approach is to take a topic from the mathematics scheme (in this case the popular Scottish Primary Mathematics group series, SPMG) and start it off in a whole-class lesson, which leads into group work with carefully-planned extension material. In this 'Timetables' topic, the activities included drawing up daily and weekly timetables and planning a family holiday which required the use of them.

Teacher: What is a timetable?
Miss, a bus timetable . . . train timetables . . . plane timetable . . .
(The children offer examples from their experience, since they are not 'empty vessels', but do not really answer her question.)
Teacher: But what does a timetable *do*?
It tells us when the bus leaves . . . when it arrives.
It tells us when to catch the train.
Teacher: Is it useful?
Of course it is, we won't miss the train if we know when it leaves.
Teacher: Do we have other timetables?
We've got to have one to know when to change lessons.
And when to use the hall.
Miss, when it's time for a TV programme.
And when it's break time!
Or dinner time!

Teacher: Are there any other timetables we use?
We have one for the school trip.
Teacher: Does each one of you have a personal, only you use it, timetable?
Yes, getting up in the morning, coming to school.

The teacher draws out other individual things the children do – 'Brownies on Tuesday', 'My jazz band on Saturday', 'Football practice on Wednesday' and so on. She talks about the presentation of timetables, what they look like, sequential happenings. Much fun arises from attempts to estimate the time for particular activities. Then the teacher constructs a timetable plan on the blackboard, which the children use as a basis of their individual timetables. Group tasks for the week from the SPMG scheme are then followed, while she rotates around the groups. Notice how her whole class session has related what otherwise might be 'just another worksheet' to the children's own experience and interests. By the end of a week, the groups will usually be at widely differing levels in their work. In this topic, for example, one group went on to plan a holiday for two adults and two children using travel bureau material. This generated much discussion and calculator work, over matters such as discounts and percentages. At a later stage (usually Friday) all groups report back, giving reasons for their choices, facts and figures, and the ways in which they arrived at their results.

Follow-up activities can often be related to previous work – for instance at a later date the teacher asked three boys who were talking about their imaginary holiday, 'How long did it take the plane to travel to Spain?' They found from their brochure that it was 2¼ hours. It had travelled 1300 miles, so their teacher asked them to estimate its speed and check by calculation. Notice once again how she has 'personalised' work on speed in her SPMG scheme. She then set them the practical activity of finding out how fast they cycled to school (but without leaving the school grounds or using a speedometer). The group had to solve some tricky problems, particularly over timing. This was eventually solved by a flag-dropping signal to start the watch being used to time the finish. The teacher made an interesting comment about reporting back on this occasion, where the group consisted of three high ability children. 'Reporting back to their ability group,' she said, 'was profitable, but to the class as a whole, especially the slower learners, was not very rewarding. There is a case for limiting reporting back, when advanced activities are involved, to peer groups and superiors, if any.'

Is a square a rectangle? – A Socratic discussion

I now want to describe a whole-class pattern which is rather more difficult to conduct. As we saw in Chaper 9, when numbers increase, firmer leadership is required, in order to maintain direction and momentum. Yet we would still like to combine the advantages of small group discussion with the sharing of ideas among the whole class. In the Socratic pattern, this is done by alternating short periods of small group activity with whole-class periods in which the ideas generated are drawn out and commented upon. The lesson proceeds through several interludes of this type, during which the teacher tries to develop the groups' ideas in a sustained way, by keeping them focussed on an 'agenda'. I usually base such an agenda around a sequence of key questions which I have planned. This probably explains the name 'Socratic discussion', which I believe was coined by Leslie Button. The flow of this lesson pattern is shown in Figure 8.2.

The following record is of a demonstration lesson with a class of fourth year juniors. It illustrates the main features of the Socratic pattern and highlights some of the on-the-spot decision making that is necessary in handling it. On this occasion the children sat in groups of three, arranged in open form so that they more or less faced a centrally-placed blackboard. It was only possible to record

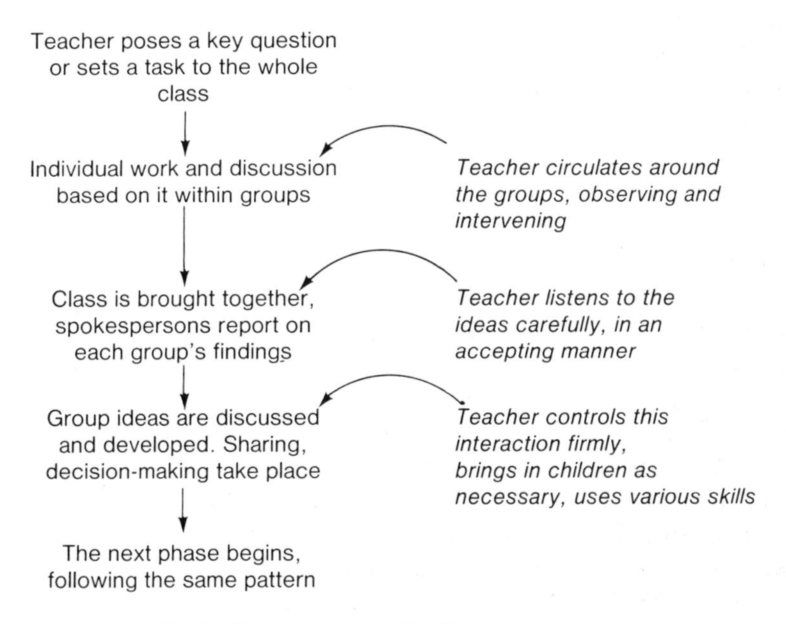

FIGURE 8.2 Socratic discussion

the whole-class interactions; a great deal of small group discussion took place in each phase. The previous lesson was of the starting point type described in the last section. It was based on the mathematical situation about square and rectangular numbers that I outlined in Chapter 6. In the course of my group discussions during that lesson it became apparent that the children, as is often the case, were using the words 'square' and 'rectangle' in an exclusive sense. The initial instruction runs 'Draw as many rectangles as you can, containing . . .', and sure enough, when they came to the case of 9, several groups asked me 'Can we draw a square, sir?' I also found that the class could use the word 'quadrilateral' meaningfully. So, in the lesson to be reported, I aimed to talk the class through to the correct mathematical usage, in which the three terms are used inclusively. That is, the square and rectangle are both types of quadrilateral (in being four-sided) and the square is a kind of rectangle (with all its sides equal in length).

The lesson began with a warming-up session using oral arithmetic. Here I used a simple idea to produce open questions – supplying an answer and inviting the children to supply different sums to go with it. Next, I used an incident from an earlier lesson, in which Janet had told me how she tried to help an eight-year-old to understand 'multiplying by nought'. I drew Janet out and asked her to explain the incident to the class. Then I asked the groups to discuss how they would have explained it. After a few minutes, during which I circulated, I brought the class together and asked spokespersons from each group to explain their ideas. This got things going, and provided some fascinating suggestions, as well as giving me information about how the children were thinking about this idea. The outcome was that it was a very difficult concept to put over – the obvious moral for readers is never to take this operation too rapidly. The idea of 'zero' as a *placeholder*, encountered in learning about place-value notation, brings about a later encounter with 'multiplication by nought' in learning about the multiplication algorithm. '3×0' can seem intelligible as 'three empty sets', or using objects, as three boxes or hoops containing no counters. But '0×3', which is also required, seems a strange idea – how can it possibly be 'no lots of three'? Here I seem first to think of a set of three objects, and then rub it out again somehow! No wonder children find it a difficult concept to explain. Here we are trying to construct a consistent framework of language, and because, for all other numbers $a \times b = b \times a$, we should *define* $0 \times 3 = 3 \times 0 = 0$. Calculators may be helpful once again!

The lesson now moved to its main theme, as follows:

Teacher: All right, that's something we'll have to come back to – I want to take up something we were doing last time. Remember we were doing all those shapes? We'll get them out in a minute, but before that I seem to remember that somebody started talking about quadrilaterals, and I think it was you, wasn't it, Robert?

(At this point, after Robert spoke briefly, the groups discussed the meaning of the word 'quadrilateral' for a minute or two.)

Teacher: Now, all this business about quadrilaterals – what is a quadrilateral?

Robert: It's a shape with four sides.

Teacher: And I think somebody in this group said they had to be straight, didn't they, when you were discussing it? . . . All right, now, Philip, could you give each group one of these sheets of paper, please? Got your pens? What I want you to do is to talk together, and draw me some quadrilaterals *but* – those quadrilaterals must have all their angles ninety degrees . . .

(Following my key question, the lesson switched to group working once again, while I circulated. After a time, judging that enough diagrams were drawn, and using my impressions from conversations within the groups, I drew the class together.)

Teacher: I want you to hold up your paper so that we can all have a look and see what you did . . . OK, put the papers down again. You've noticed something, I expect, haven't you?

(The sets of shapes drawn are all very similar in every group, of course.)

Teacher: Now, let's go round again, who's going to talk for this group this time? I've forgotten, was it Sally – Philip – yes?

Philip: We only found some squares and some rectangles.

Teacher: That's all you were able to find, was it?

Philip: Yes.

Sally (interjecting): We did small rectangles and squares.

Teacher: Ah, you could draw them different sizes, could you?

Sally: Yes.

Teacher: Hmm, that's interesting. What about this group, now? You did the talking last time, Bill, didn't you? Robert? You started all this with this quadrilateral business, now what about your group? Could you talk to the class, by the way, and not to me?

Robert: We found the same as their group did but we found out that if you put this square at a different angle it looks a bit like a rect- a diamond.

Teacher: Can you just explain that a bit more? Do you want to hold that paper up and point, or draw another one, Robert? . . . Which one was that? Was it in your first drawings or a later one?

Robert: This was one square – and these four – and then we had one like this –

Teacher: I don't think any other group did this, did they?

Robert: We found out the diamond shape was still a square.

Teacher: The diamond shape was still a square. OK, thank you very much, Robert. What about this group? You spoke last time – OK, Lisa, what did your group find out?

(Lisa, a very shy pupil, froze and was unable to reply.)

Lisa (whispers almost inaudibly): The same.

Teacher: Well, no, come on, you've got to say.

(There was a pause, in which I moved up to support Lisa, and spoke quietly and gently to her, temporarily ignoring the rest of the class.)

Teacher: Lisa . . . What did you find out, Lisa – can you point to what you found out? (She does so.) Now tell me about those shapes, Lisa, please – how many did you find? (The last very softly to Lisa only.)

Lisa: Five.

Teacher: Five, was it? What kinds of shape was it?

Lisa: Diamonds, squares and rectangles.

Teacher: Can you all hear, by the way, what we're

	saying here? Can you speak a bit more loudly, Lisa?
Lisa (speaking more clearly and audibly):	We found rectangles.
Teacher:	You found rectangles, did you?
Lisa:	Squares and diamonds.
Teacher:	Squares and diamonds? It sounds a little bit like this group did. Which one was your diamond, Lisa? Can you show us the one you had? . . . Can you just show it so that Elizabeth can see as well? . . . OK, who spoke for this group last time? . . . Your turn, Mary?

Before I continue, this incident with the shy, inarticulate Lisa is worth thinking about. I am not much attracted to the idea of group leaders, although Easley, whose work I shall report later, strongly advocates their use. All our children, as I shall describe more fully in Part Four, need to develop their oral powers to the greatest extent they can. How can we seriously claim to be developing the social skills mentioned in Chapter 1, unless we try to do this with all? We must not rely on the articulate, clever or dominant children to carry the discussions forward, but try to draw everyone out, even shy or inarticulate children like Lisa. The other children in the class can be encouraged to be supportive, just as the teacher should be in acting as a role model. As Button has argued, why should supportive behaviour like praise, encouragement, or simply patience, be the prerogative only of the teacher's role? Shy or inarticulate children can be encouraged to play a greater part by turn-taking games such as I have already described in earlier chapters. But they also need opportunities to present group ideas in public to the whole class. That is why I persist with Lisa in the present case, instead of taking an obvious way out. (She was under extra pressure on this occasion, remember, because of a watching audience of teachers.)

The teacher may need to take a very firm line over these matters, not only insisting on turn-taking, but on other children providing a patient and attentive audience. Note also the aside to Robert, a little earlier, about talking to the class and not to the teacher. This firmly-ingrained habit, resulting from the traditional three-term pattern of interaction (Chapter 2), needs to be discouraged. As a general rule, the talk in the whole class phases should be public and clearly audible to everyone. Statements should be repeated, if necessary, as in Lisa's case once she had gained a little confidence. My statement beginning 'Can you all hear, by the way . . .' was partly to get Lisa to repeat, partly to quell a murmur of group discussion that was breaking out. Let's see how the Socratic discussion continued!

Mary: Yes. We found ten quadrilaterals and also found that it doesn't really matter how long the sides are, they have all got ninety degrees.

Teacher: Hmm . . . what kind of shapes did you find?

Mary: We found roughly the same as Robert's group over there – we found the diamonds, rectangles and squares.

Teacher: Hmm, people have talked a lot about these diamonds, haven't they? But you said something interesting, didn't you, Robert, about that? What did you say? Can you remind us?

Robert: It's a diamond shape but – it looks like a diamond shape but it's still a square.

Teacher: What do you think about that, Mary?

Mary: Well, I've been sticking to it about forty-five degrees and looking at it and it's just like a square.

I had not foreseen this appearance of diamonds everywhere! You can see how the children had cleverly invented a fresh shape to satisfy my opening condition, that I had not thought of myself! The conclusions that emerged from this important whole-class phase were that my original instruction had compelled the children to draw diamonds, squares, rectangles and oblongs, that squares were a kind of diamond and that there was another kind of diamond besides squares (not belonging to this set of shapes). Robert made a particularly lengthy contribution which included the statement 'If you move the square to its side it looks like a diamond but all its sides are forty-five degrees and there's another diamond turned on its side to look like forty-five degrees'. I moved the class into the next phase of groupwork with another (planned) key question, 'All those shapes you've been drawing for me *I call rectangles*. So you've been drawing two kinds of rectangles. Now, two minutes to talk – *what do I mean by that*?'

The usual buzz of group discussion ensued, ended by my asking 'Now what have we decided?' and calling different spokespersons to report as usual. The crucial statement, which came from Peter, caused general amusement. I had pigeon-holed an important discussion I had in his group, in which 'oblongs' were introduced – 'A baby word!' said Peter scathingly. 'I use it!' I retorted. So when he reported 'There are two kinds of rectangle, a square and er – an oblong!' his tone was one of reluctant apology at being compelled to use this baby word. The matter had spread as an ongoing little joke around the class, deliberately encouraged by me, of course. The final group activity in the Socratic session was to make drawings of an 'oblong rectangle' and a 'square rectangle'. After this the lesson continued with the squared paper activity begun in the previous lesson, developing the mathematical situation already described in Chapter 6.

Figure 8.3 shows some of the drawings produced during the Socratic part of the lesson. I certainly felt during it that negotiation of meaning was taking place, in the general direction which I had planned. Robert's and Mary's 'forty-five degrees' meant, I think, a square 'tilted up' through that angle (approximately) from an imaginary reference line drawn across their page. The fact that the groups are using the same sheet of paper can sometimes force these changes of viewpoint. Robert's 'other kind of diamond' is properly a rhombus, of course, so I have a starting point for yet another discussion at a later date! I used the discussion about multiplying by nought with Janet in a similar way to start off this lesson.

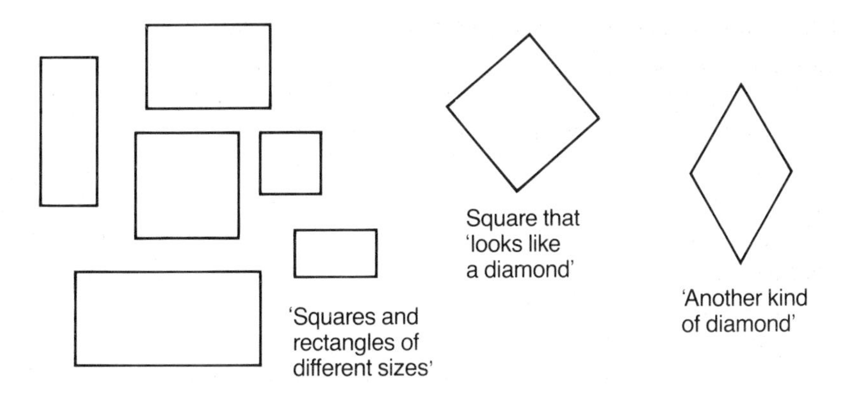

FIGURE 8.3

Planning for Socratic discussion

In Socratic discussion, each time that a key question is asked by the teacher, we break out of the three-term sequence by immediately providing the children with a 'thinking opportunity' and a 'discussion opportunity', instead of demanding responses on the spot. The teacher circulates round the groups, watching or listening unobtrusively, or intervening using the skills that have been developed in Part Two for small groups. When she or he judges that sufficient ideas are available, and has gained information about how the children are thinking (for instance, about any difficulties that have emerged), the groups are called together as indicated in the extract. The teacher uses the ideas to lead into the next stage, injecting perhaps some explanation and language before posing the next key question.

The advantages of this pattern are obvious. When the class is brought together the teacher should already have a good picture of the kinds of responses the key question has provoked, and can act accordingly. Children are drawn out, and responses extended, from prior knowledge. One group's ideas can be used to influence the others, or conflicting views can be brought into contact and argued out – a small scale version of Lakatos' 'proofs and refutations'. I feel that the teacher's position is very much stronger than in conventional question and answer, where questions are fired off into the blue, as it were. Moreover, the children have already been able to talk together about their ideas and difficulties. Spokespersons are likely to gain confidence because they speak for the group, rather than themselves, and can call upon other members for support, if necessary. Also, the teacher may already have talked with them in their group, and is, in effect, inviting them to share what was said with the class as a whole.

Teachers should find that their children will come to expect these thinking and discussion opportunities, instead of expecting to compete for the teacher's attention at once, as in the patterns described in Chapter 2. You are 'rewriting your mathematics script'! Socratic discussion can proceed through several phases, but would normally lead into a more extended group activity of the usual kind. A discussion of about half an hour could have four or five phases, each lasting several minutes. Sufficient time should be allowed in the groupwork for thinking and discussion, but it should be broken off *while groups are still active* – this may need firmness, but keeps the momentum of the lesson going. As long as enough ideas are available, it does not matter whether or not everyone reaches some final stage, since comparing, sharing and discussing forms the theme of the whole-class phase that follows. It is important not to let a group run out of ideas and come to a standstill – if needs be, introduce material from another, more productive group. Thus as well as circulating, it is necessary to keep a general overview of all the discussions.

Seating also needs some careful attention, and there are some extra requirements in the case of whole class lessons. It is important that children in groups face one another, rather than sitting in a row.

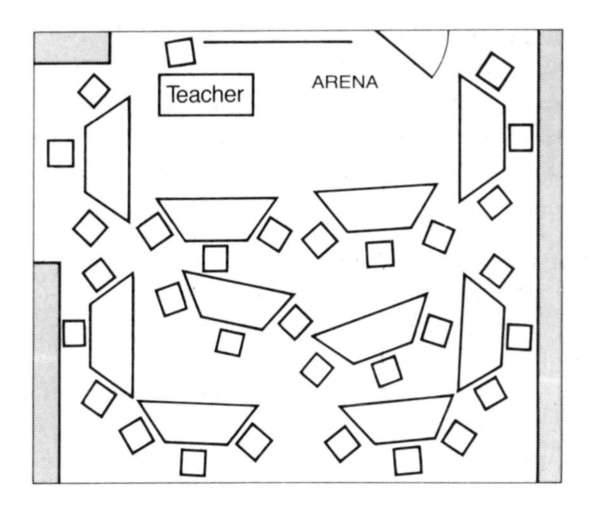

FIGURE 8.4

This takes care of the first level of interaction, between the children themselves. But our groups also need to collaborate, and so there will need to be some central focus, or arena, usually around the blackboard, so that the second level of sharing can take place. An ideal arrangement, not always possible to realise completely, is shown in Figure 8.4. Here each group is open, rather like a horseshoe, with its open side facing the arena. The horseshoes are placed so that not too much craning around is required when spokespersons report to the whole class, and children can move to the blackboard if necessary to demonstrate their ideas.

With the groups allocated, probably with mixed ability, and their seating organised as far as is possible within the constraints of a particular classroom, the teacher can keep track of events most simply by thinking of the class as consisting of a rather small number of groups, rather than thirty or so individuals. For example, you could work with ten groups of three. I find that three is an ideal size for a group, to keep everyone involved, but the actual number of groups can then become rather large. Alternatively, you might prefer to cope with a smaller number of rather larger groups, say six groups of four or five children, until confidence is gained. There may well be hangers-on, or some splintering, if you use these larger groups, and this may need watching.

The work of Easley, in the United States, is based on ideas related to those expounded here. Easley believes that primary teachers should leave most discussion of mathematical and physical science content to their pupils. Cognitive research shows that young children develop and test alternative rational explanations which authoritative explanations cannot displace. Conflicts arise between these and the pupils' own unexamined concepts which can generate severe anxieties about these subjects. Easley advocates an approach based on peer group discussions, which teachers supervise 'from one level up'. These discussions are to be based on interesting and challenging tasks which will occupy pupils for about an hour. Group leaders can be trained to call on each group member for an opinion; other techniques include polling the class, ensuring that every child's view is counted. Easley writes (1984):

> I believe that . . . primary teachers should moderate and support classroom communication in mathematics and science, and not rule on what is right and wrong. They should set an example for group leaders by providing an atmosphere in which ideas can be expressed without fear of being put down, thus providing safe outlets for the emotional concomitants of pupil thinking. Even if all pupils do not reach a consensus on the problem or task at hand, they are gaining in courage to attack unfamiliar problems.

Easley's ideas are based on lengthy field work studies, including a period spent in Japan at the Kitamaeno School in Tokyo, where he observed teachers using peer group dialogues to teach much of the mathematics and science. Here mathematics achievement was 'very high by American standards'. Easley's approach may be based partly on a feeling that primary teachers (in the USA) do not possess the large repertoires of technical ideas and expertise available to specialists. The teachers may see the advantages in demonstrations by such specialists, but could not imitate them except in very small steps. Easley's programme has come to be called 'learning by listening towards'.

The procedural role in Socratic discussion

Handling a lesson of this kind clearly demands a number of teaching skills. The basis for these is best developed in the safer context of small group work, where I have already introduced the terminology of drawing out and extending children's responses. In general, the teacher is trying to minimise her or his own mathematical activity, so as to get the children to do as much of the thinking as possible. An effective mathematical situation should facilitate this, as we have seen, and enable the children to build and test their ideas. The teacher still has quite a lot to do, however, in keeping things moving constructively, and particularly in a whole-class session of the Socratic type. In Chapter 3 I introduced the term 'procedural role' to describe this type of intervention. In this section I shall list the basic techniques that I have found helpful, illustrating some of them briefly with reference to the extract just given.

It seems to me that the teacher is effectively doing two distinct things at once – trying to follow the mathematical thinking of the children and also using procedural interventions to keep the class working constructively. It is obviously of fundamental importance to plan an effective mathematical situation and let it work for you, thus minimising the mathematics you need to do. After that, you are trying to 'rewrite the mathematics script' so that the expectations and patterns in the procedural role become matters of habit, for you and the children. The various skills that I summarise below will thus be exercised more or less subconsciously, so that you can afford to listen in a more relaxed way to the children's ideas.

1 *Avoid the three-term sequence* Accept responses from the children, as, for example, in the extract, 'Hmm, that's interesting', but do not feel that it is your job to evaluate them every time.
2 *Be an acute observer and a good listener* This is just as important here as it is in small group work of course, and there is likely to be much more to observe and listen to going on!
3 *Circulate round your groups* This needs to be done so that you visit every group at some stage, though not necessarily in every groupwork interlude. Also ensure that every group's ideas are brought in at some point, however briefly – try not to leave some group feeling that they have been left out. You may sometimes get the report 'We found the same as they did', particularly from someone reporting last. It is best to insist on a reasonably extended report nevertheless – repetition is unlikely to do any harm!
4 *Pigeon-hole important ideas* This means making a mental note of anything significant that occurs, either in your circulation or in the whole class interludes. Use this mental reference file to draw out pupils at appropriate points – for example as I did with Robert in the extract, over his idea about the diamonds. The file is for future reference, rather than immediate action – I might note a child who will need attention later, rather than giving it at the time, when it might disturb the flow of the lesson. This kind of action I call 'tying up loose ends'.
5 *Draw out pupils' responses* This means getting responses from

individuals or groups, either by accepting offers, or calling on children on the basis of information which you have pigeon-holed. Other techniques include pausing, 'keeping the gate open' for further ideas. For example, in Chapter 4 in the *1, 2 and 3* extract, the teacher continues to look for ideas even when the children have found all six of the arrangements - she allows them to decide that six is all there are to find. 'Are there any more ideas?' is a straightforward question to ask.

6 *Extend pupils' responses* This means getting more information from an individual or group - numerous examples of this are to be found in my accounts. 'How did you get that?' or 'What do you mean?' are useful questions to extend. Notice how hard I have to work to get Lisa to extend her response, compared with Philip, Sally or Robert.

7 *Use deflecting* This is a more subtle technique needed if you do not wish to respond to a demand for you to evaluate an idea. Instances are hinted at in the *Ergo* extracts in Chapter 5, but more clearly in the Socratic discussion, when I say 'What do you think about that, Mary?' after a statement by Robert during which he looked intently at me expecting some comment. This form of question is often useful as a deflecting strategy, not necessarily aimed at a particular person. The teacher suggests, directly or indirectly, that someone in the class comment on a response initially directed at the teacher.

8 *Develop linking* This is another subtle technique, in which the teacher suggests that the current response be compared with another that has been put forward, or that the teacher knows about from discussion in another group. It is a way of ensuring that ideas from different people are compared or that children are put in touch with one another's ideas. Since the teacher is in a dominant position with an overview, unlike the children, this technique is of considerable importance. It is well worth developing, as it can greatly improve the quality of a lesson. It is clearly dependent on acquiring skill in pigeon-holing. In the extract, an obvious example occurs when I say 'But you said something interesting, didn't you, Robert, about that? . . . Can you remind us?' The class is put in touch with an idea that only I, and Robert's group, know about at that stage.

9 *Keep an overview* Every teacher needs this skill every day - keeping a general eye on what is happening, while at the same time handling other matters. It may be related to, or identical with, the skill which the researcher Kounin called 'overlapping', the ability to handle two matters at once. He found it vital in effective class control.

10 *Use scanning* This is another important teaching skill, based on maintaining general eye-contact so as to spot pupils ready to contribute (in this case). It is also vital, like overviewing, in class control.

Comparing the traditional and rewritten scripts in whole class work

Below, to conclude this chapter on whole class methods, I have summarised the main features of the traditional mathematics 'script', as they emerged in Chapter 2, and contrasted them with the features of a 'rewritten script' based on Socratic discussion. The contrast may highlight more clearly what a discussion approach implies – what it is that we are trying to achieve. It is clear that it is the use of groups as the basic unit of teaching and learning that will bring about the necessary changes. Approaches other than Socratic discussion may well be tried, but this feature remains essential.

TRADITIONAL SCRIPT (Three-term sequence)	REWRITTEN SCRIPT (Socratic discussion)
Teacher's role	*Teacher's role*
• asks closed questions	• asks general questions
• makes decisions	• gets pupils to make decisions
• comments on pupils's statements, accepting or rejecting in terms of an ideal response	• rarely comments on pupil statements and never rejects
• controls the interaction	• controls only the interaction between groups
• controls the course of the work	• allows a mathematical situation to guide the course of the work
Pupils' role	*Pupils' role*
• speak to and listen to teacher	• speak to and listen to one another as well as the teacher
• few pupils take part in the interaction	• virtually all pupils take part in group or whole class interaction
• responses are matched to an ideal, rather than used	• responses are used to further the work

• responses tend to be brief	• responses are extended, particularly within groups
• often try to guess the teacher's ideal response	• think about and comment on what has actually been said
• exert little influence on the course of the work	• exert a considerable influence on the course of the work

PART FOUR
SETTING IN CONTEXT

9 Mathematics and school language policy

Communication in the classroom

Communication in the classroom is the theme of an important article by Mercer, to which I have already referred in Chapters 6 and 7. I shall summarise some of his ideas in order to broaden our discussion in the manner I indicated at the end of Chapter 7. Mercer focusses on classroom talk as the main theme of his article, and on the importance of making use of the experience and expectations which children bring to school, and the ways in which they make sense of what they do there. One of his examples is to do with listening skills, a theme that has arisen several times in the course of my discussions. Some children do retain information, Mercer says, while those of normal intelligence who do not are said to lack such skills. He suggests that the explanation for such communication failures may often lie elsewhere. Successful retention and comprehension by children

> depends very largely on such matters as (1) the motivation of both teacher and child to attend to and communicate with each other, and (2) the extent to which teacher and child have a shared understanding, a common framework of language and concepts whereby new information can be related to matters already understood and remembered.

Note the emphasis on *teacher and child* attending to one another in the quotation! A necessary basis for successful communication is the development of such a body of shared experience. In mathematics this involves, I think, the process of negotiation of meaning I described in Chapter 1.

Mercer goes on to discuss the patterns of interaction commonly found in the primary classroom, much as I set out for mathematics in Chapter 2. The overall effect of the dominant patterns is to tailor the pupils' language to the expectations of the teacher, rather than drawing out their own and refining it. It also limits the opportunities for children to develop a broader range of communication skills, such as asking effective questions, presenting one's ideas clearly, or commenting constructively on statements made by other people. So a 'school language' is set up which is distinct from, and overlaps only slightly, the 'home/out of school language' of the children, as shown in Figure 9.1.

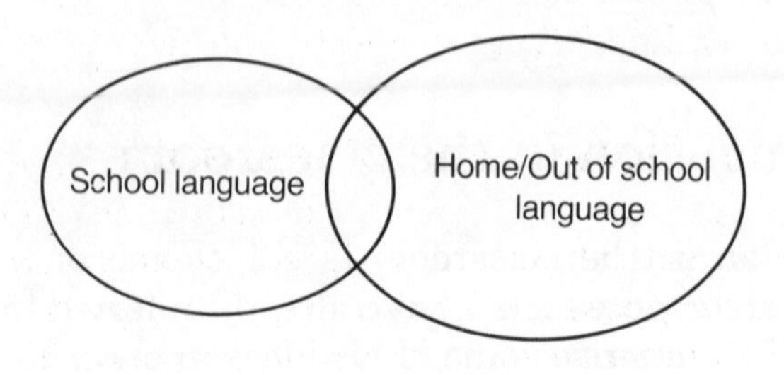

FIGURE 9.1

Wells' research (1984) into the language development of young children showed important differences between parent-child interactions and teacher-child interactions. Children in school have very limited opportunities to initiate, to ask questions, compared with at home. You may be reminded of the findings by Tann which I quoted in Chapter 7. Wells suggests that teachers should try to develop more exploratory, varied patterns of talk so that teaching and learning becomes more of a mutual activity. This pattern was hard to find in his classroom recordings, but more common in the home. Wells offers various suggestions, such as creating imaginary stories, or giving personal responses to poems, so that such patterns might develop. In a later, interesting collection, edited by Wells and Nicholls (1985), various contributors develop a whole range of ideas and suggestions around this theme. Phillips' article, on *Discourse development after the age of nine*, seems to me of particular interest, because it relates to the later stages of schooling.

He begins by reviewing the research into language development carried out in the 1970s. It shows, he suggests, that sensitive adults who are ready to listen to children carefully, and to respond flexibly and with interest, perform an irreplaceable role in the process by which young children learn to use talk for the range of functions which Halliday identified (cf Chapter 1). As children grow older, more and more material enters the school curriculum, and teachers have less time to develop conversational initiatives – 'It will never be as easy (for them) as it was for their colleagues in earlier years'. Phillips goes on to dismiss 'whole class discussion' as a major way forward. Barnes' 'Transmission – Interpretation' model appears to offer the possibility, after transmission of information has occurred in some way, of a period of interpretation in which children talk and discuss with the teacher. But sensitive listening, although helpful, is not enough, and there is not unlimited time available to negotiate. Hence some form of teacher constraint on children's linguistic freedom, however implicit, is inevitable, it seems, in teacher-led situations.

Phillips' way forward is to use peer groups to promote language development, but *not by simply allowing children to talk*. Permitting children to chatter or discuss in small groups, although superficially a response to the demand to give talk greater recognition, distracts attention from consideration of the quality and appropriateness of the talk. He describes a detailed model of peer group conversations, based on his own transcripts. This involves five modes of discourse, the Hypothetical, the Experiential, the Argumentational, the Operational and the Expositional. In each mode, the discourse contains characteristic 'markers' which signal the speaker's intention to the other members of the group. For example, in the Experiential mode, speakers begin statements with 'I remember once . . .' or 'It reminds me of when . . .' before continuing with an anecdote. There is a mutual acknowledgement, through these markers, that the discourse is to be about personal experience. Phillips goes on to link these modes with mental processes in a careful way. Neither the Operational nor the Argumentational modes, as described by him, are likely to generate higher order cognitive thought. (In case this may sound surprising, I suggest that readers consult the original article.) In the Argumentational mode, for instance, speakers are preoccupied with asserting their own views, but there is no public consideration of the reasoning by which one alternative is selected rather than another – no comparison of pros and cons.

Phillips is led to his Hypothetical and Experiential modes as those offering most potential for development. Both modes oblige members to review the conversation itself. He makes various suggestions about possible forms of teacher intervention, which would leave groups free to continue, but make them examine their own discourse more carefully. In a practical activity, for instance, there could be a planning stage, then an activity stage, followed by a withdrawal for further discussion and reconsideration/

modification of ideas, before a final report stage. This could be established as a routine for solving a mathematical problem, or carrying out a science experiment. Discussion topics could be framed in ways which invited speculation and left conclusions open, rather than requiring decisions to be made.

It is clear both from this account and from Mercer's original article that there is wide support, founded on both practice and research, for the view that modifying established patterns of communication in the classroom can assist children's learning in very significant ways. This applies, not only to younger children at the infant stage, but also to children in the junior school and later. Moreover, according to Phillips, the answer to the question I raised at the end of Chapter 7 ('Should we let nature takes its course and let pupils develop the skills of cooperation the best way they can?') is a definite 'No'. Some forms of intervention are desirable in order to improve the quality of the discourse in peer groups. Mathematics is a comparatively late arrival on this scene but it could make a substantial contribution to solving the difficulties outlined by Phillips. For example, the use of investigations I outlined in Chapter 6 could help pupils to focus on the Hypothetical mode he identifies, as well as on the planning approach described above. Skemp's building and testing framework would place peer group discussion in a setting which could sharpen both the Operational and Argumentational modes which Phillips appears to play down somewhat. Groups might need to be encouraged to look at their own performance as a group, as well as at the mathematics which they have achieved. Here the cassette recordings I have mentioned in earlier chapters might have another part to play, by offering feedback to the participants. Or it might be preferable to keep interventions in a lower key – suggesting to a group that more ideas are needed before proceeding into an activity, for instance.

The teacher is definitely teaching, in either case – but it will be necessary for her or him to try to identify the aims and processes involved more clearly than is the case at present. The potential clash or mismatch between teacher's and children's perceptions appears once again here. Mercer discusses it at some length, as in my earlier references. It seems to me that it might be best, as children move into the junior stage, to discuss these matters more and more openly, so that there is less room for misunderstanding on the part of the children about the aims of their group activities. They might also find the build and test idea helpful and appealing – I wonder if it might form part of the teaching, put in a

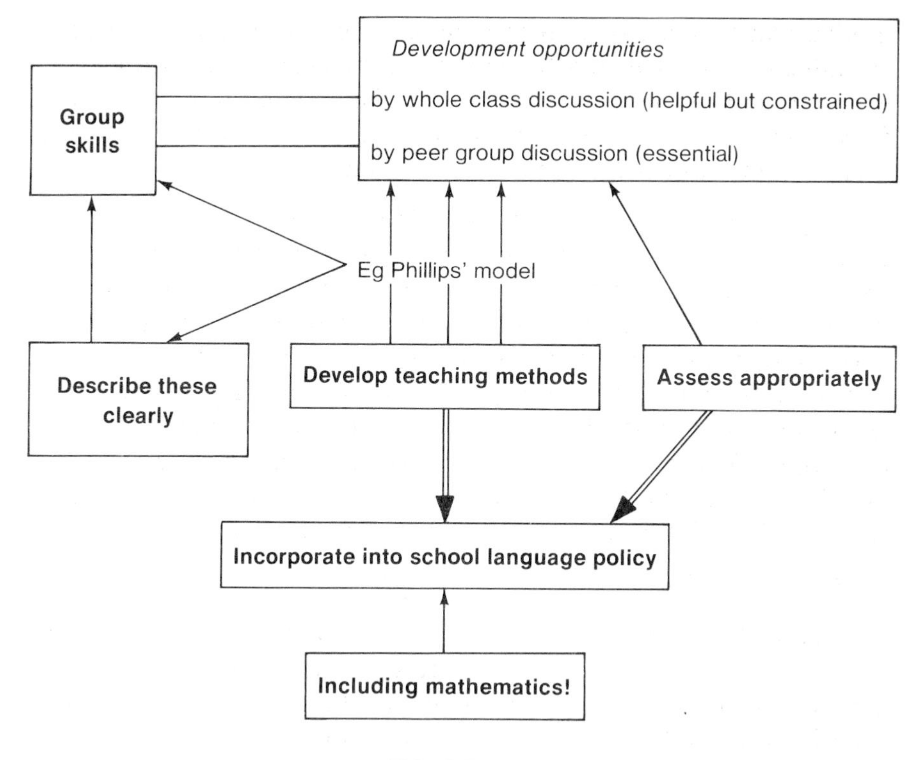

FIGURE 9.2

simple way? Finally, if we think we are teaching something, however elusive, and anticipate progress, then assessment in some form has to be considered. Readers at this point may feel that the discussion has been somewhat tortuous, so I shall try to bring together the various aspects in Figure 9.2.

Mercer suggests in his article that theories of cognitive development have been too individualistic. This development is now thought to be tied up with social development, so that communication between individuals, especially during problem solving, is of great significance. 'Translated into classroom practice, this means (amongst other things) setting up more structures which enable children to communicate with one another and to practise and develop their communicative abilities.' For this purpose he offers the rationale set up by Silvaine Wiles (1981) to do with 'communicative competence'. The Lakatos group found this framework a helpful one – particularly as they lacked the time and expertise to set up their own! – and tried to link up ideas from mathematics with those put forward by Wiles. The idea seems at least a start, for schools interested in constructing a language policy which incorporates mathematics. Moreover, it seemed a good idea to try to put our own in-service suggestions into the frame of another in-service training pack on the language theme.

Communicative competence and mathematics

Wiles' framework is based on the idea that all language (reading, writing, speaking, listening) involves communication of some kind, and so it is vital to look at children's competence in this area. She lists a set of essential aspects of 'communicative competence':

- the ability to listen and make sense of information;
- the ability to take another person's view into account, modifying one's approach if necessary;
- the ability to make contributions appropriate to a particular audience (friend, teacher, younger child etc);
- the ability to reformulate contributions where necessary (repetition, self-correction, putting it another way);
- the ability to recognise and adhere to discussion conventions, such as taking one's turn, acknowledging the contributions of others, understanding the techniques for changing tack or bringing conversation to a close.

This in fact is a formidable list, on which many adults might fail in various aspects. Nevertheless, it is a good start, and one can see where many of the details I have mentioned in earlier chapters fall into place. Wiles goes on to discuss the benefits of small group work and to list some features of the teacher's role. It is interesting to compare these with points which I have highlighted. For example, Wiles says that the teacher needs 'to set up groups that will work well together' and 'spend enough time with each group to establish successful working patterns'. These are matters that I discussed in Chapter 7 in covering organisation. 'Setting clearly defined tasks' is important – but as I have said, the aims and perceptions of teacher and children need to be taken into account. Wiles says that the teacher should 'dip into the group activity' not only as a tester but also to participate and learn from the children's discussion. I hope that I have by now treated this point in enough detail for its application in mathematics to be seen. Finally she says that the teacher needs to demonstrate the value placed on learning through talking by providing an audience for the group's activities at times. This can be done by oral presentations at the conclusion of an activity, by displays (the most common method), and also by the use of Socratic discussion with the whole class, as I have described in Chapter 8.

What then can mathematics contribute to Wiles' 'communicative competence'? I listed the following as potential, and actual if the programme developed in this book is realised:

- articulating and presenting an idea publicly, in a clear and intelligible way;
- explaining a method;
- arguing logically in support of an idea;
- criticising an argument logically, including one's own;
- evaluating the correctness of an idea, or its potential in attacking a problem;
- speculating, conjecturing, entertaining an idea provisionally;
- accepting an idea provisionally and examing the consequences;
- keeping track of a discussion, reviewing;
- coping with being stuck, supporting others in difficulty;
- drawing others out, using 'Show me . . .' or 'How did you get that . . .?'!
- acting as spokesperson for a group's ideas.

When I looked through this list, listening, which has figured so often, was particularly prominent as a skill required in many of these aspects. Although I do not see any of them as specific to mathematics, they seem in many cases very sharply exemplified in it. For example, the ideas in mathematics are likely to be particularly difficult to articulate clearly. Later, the 'What if . . .?' and 'What if not . . .?' aspects of mathematical activity could be particularly useful in the Hypothetical mode, as a 'playing around with ideas'. Careful listening will be required throughout – just as, in Parts One and Two, I stressed that the teacher needs to become 'a good listener and an acute observer', so the children will need to acquire the corresponding skills themselves. The teacher needs to find out about the children's mathematical thinking and they need to find out about one another's. Evidently the teacher will need to act as a good role model in this respect!

In trying to integrate mathematics into a language policy, staff might proceed by examining the statements already made concerning such policies, either in more general contexts – such as I have been quoting – or in other curriculum areas, notably English. The Bullock Report, *A Language for Life* (1975) ought obviously to be a source of such ideas. It recommended training in 'ear language' for teachers, for example. But Hawkins (1987) criticises the Report sharply and blames the comparative failure of its proposals on its muddled thinking and contradictory recommendations. About listening ability, for example, Hawkins points out that on the one hand the Report states 'In our view the ability (of listening) can best be developed as part of the normal work of the classroom and in association with other learning experiences', but later, 'deliberate strategies may be required for it cannot be assumed that the improvement (in listening ability) will take place automatically'. Hawkins himself is in no doubt about this particular matter and gives a detailed programme for 'Learning to listen' in an appendix. Farrar Kinder, in a paper presented to the Sixth Annual Conference of UKRA (1969), thinks that improvements in children's listening support improvement in other language skills, such as generalising information read, or including more vivid and accurate detail in their writing.

One school started by looking at the objectives relating to speech in *English from 5 to 16* (1984), a parallel booklet to the mathematics one which I have used so freely. Some of these objectives are as follows:

Objectives in speech at ages 7, 11, and 16

Age 7	*Age 11*	*Age 16*
Speak clearly and audibly	Speak clearly and audibly	Speak clearly, audibly and pleasantly
Narrate simple experience and events	Participate courteously and constructively in discussion	Describe experiences clearly
Discuss constructively a group task	Make clear statements of fact	Clearly explain a process
Ask relevant questions	Frame pertinent questions	Argue a case

These statements do not form the full list, nor do I wish to go into whether they are appropriate objectives for the ages stated. Provisionally accepting them, it is easy to see how they would all be exemplified in the kind of mathematics curriculum that has been advocated in the previous chapters. Language cannot be regarded as the prerogative of the English curriculum, and nor can imagination or creativity. I have noted that proposals to help teachers adopt more open approaches (such as those offered by

Wells and Phillips which I quoted earlier) appear to assume that mathematics (and probably science) have little to offer in the latter respect. Poetry, story-telling, painting, writing, are laid before us - but why not mathematical investigations?

What of the multiethnic classroom, in which English may be a second language for many children, who speak a variety of first languages such as Punjabi, Urdu and many others? Wiles writes at some length about this, in the reference already cited (Wells and Nicholls). Today, she argues, educators are more likely to see bilingualism as a resource, not a handicap. The regular classroom is the best context for second language learning, for a variety of reasons which she describes in detail. Here, Wiles asserts,

> a second language student is with a group of pupils of her/his own age . . . (and) will learn most effectively in the mainstream class. Any additional support should be given to bilingual children in the regular class in curriculum-related areas in conjunction with the class/subject teacher.

The reasons discussed by Wiles are all related to the use of peer group talk. The approach to mathematics presented in this book could therefore offer advantages in the multiethnic context. Our own very modest experience, involving first-language English speakers in a Welsh medium primary school, also offers support for the view put forward by Wiles. It is to be hoped that teachers working in multiethnic classrooms will begin to build up, and publish, a body of informed expertise on the teaching of mathematics based on use of talk, as it applies in their own context.

Assessment

It is appropriate to say something about assessment at this point. In Chapter 1 I listed four groups of reasons for the use of talk in teaching and learning mathematics. Among these, its use as a *means of assessing* the other three (developing understanding, improving language and developing social skills) played a kind of unifying role, as shown in Figure 1.1. As far as mathematics is concerned, we are initially interested in the understanding and in the use of language in relation to mathematics. But in fitting mathematics into a more general picture, we may become interested in assessing the social skills and the general quality of talk, as they are observed in the mathematics lesson as well as in

English or other areas of the curriculum. For example, we may wish to assess progress in terms of Phillips' 'discourse development' in some way, or to measure the extent to which the objectives listed under speech in *English 5–16* are developing in our mathematics activities.

As I suggested in my opening section, I think this attempt is dependent on a clear identification of a teaching programme and of the objectives involved. Silvaine Wiles' 'communicative competence' offers a starting point for our aims, and the list which I gave in the same section is very close to an example of a set of possible related process objectives for mathematics.

Tough (1976) offers some frameworks for recording characteristics of speech, such as

1 How often do children use speech to draw attention to their own needs or to maintain their own status by defence or assertion?
2 How often do they use language to extend or promote action and to secure collaboration?

A possible method of assessment might be a checklist based on a list of characteristics of this kind. Examples are offered by the Schools Council (1981) in a helpful reference. Checklists are basically an aid to teachers in making and recording their observations of children's development in different areas of the curriculum. They probably need to be developed by the staff through discussion so as to ensure a shared understanding of what is being observed and assessed. Some characteristics might be specific to a given curriculum area such as mathematics, others might apply much more widely. Each characteristic is linked with a scale, which might be a very simple one of the 'true/not true' type, or have several scale points with the extremes and means defined by an attribute. For example, the characteristic of 'curiosity' in a checklist given in *Learning through science* is carefully defined as follows:

CURIOSITY

Often seems unaware of new things and shows little sign of interest even when they are pointed out.	1
Is attracted by new things but looks at them only superficially or for a short time; asks questions mostly about what things are or where they come from rather than about how or why they work or relate to other things.	2
	3
	4
Shows interest in new or unusual things and notices details. Seeks by questioning or action to find out about and to explain causes and relationships.	5

The top and bottom statements define the ends of the scale, and the middle statement the centre of it, with two other scale values in between. A three point scale would be a simpler compromise between this and the basic two point one. One primary school has developed checklists to assess its wider language aims, using a simple 'largely true/largely not true' scale. One of these is shown below in Figure 9.3. It is clear that such checklists could be used as part of the mathematics assessment, but are not specific to it. Much development work is going on in this area of appraisal and assessment, and no doubt further information will rapidly become available to primary staff. But it will always be important to ensure that teachers understand the characteristics they are observing. Another good reason for becoming 'good listeners and acute observers'!

Assessing wider aims

Assessing the group	Largely true	Largely not true
Do the children talk to each other about their work?		
Can they listen to each other?		
How often do they use language to extend or promote action and to secure collaboration?		
Do they use language to direct action of other members of the group?		
Do the children share their ideas?		
Are they articulate in their attempt to convey meaning?		
Do the children compare their ideas with one another?		
Can the children deal with differences in results on a reasonably objective basis, not being completely swayed by considerations of others?		
Do the children challenge ideas and interpretations with the purpose of reaching a deeper understanding?		
Do the children recognise conflicting evidence or conflicting points of view?		
How often do the children use language to reason, to recognise and offer solutions to problems, to contemplate alternatives in real or imagined contexts?		

FIGURE 9.3

Enlisting parental cooperation

A primary school has several interested groups to satisfy, when it tries to develop a new approach to the curriculum. The Inspectorate, the local authority and its advisory staff, the school governors, the secondary school to which it normally 'feeds' its pupils, the parents of its pupils . . . the list is formidable, and it is possible to speculate about priorities. After glancing briefly at the secondary school, I shall focus on the question of enlisting parental cooperation in this final section of the chapter.

Several years ago I reported on some discussions among heads of secondary mathematics departments, concerning relationships with primary feeder schools (Brissenden, 1984). What influence, if any, should the department have over methods of teaching mathematics in their feeder schools? Should the departmental policy be an agreed one, which draws in the primary teachers' own views and ideas? The guidelines issued by the West Glamorgan authority (Gray, 1979), which were based on such consultations, are a typical way of trying to ensure continuity across the changeover at eleven. At the time of the discussions I reported, there was great variety among secondary schools over methods of creating these links, by exchange visits, visits by children and so on. Some did very little, while others tried quite hard. The article by Holmes, for example, in the reference just mentioned, shows a humane head of department who, in writing of the need for close liaison, says 'How can those very same pupils, three months after leaving their primary, be expected to change suddenly and survive without the aid of practical work?' An abrupt change in methods could clearly have a shattering effect. The Inspectorate were critical of the manner in which secondary schools focussed on assessment for the purpose of placement, rather than diagnosis, in the first year of secondary school. What use do (and should) secondary schools make of the information about children's mathematical attainment supplied by the primary school? All too often very little, I suspect, in their concern with allocating pupils to the 'right set'.

An abrupt change from a DOING, TALKING, RECORDING approach using groupwork to a more formal one based on 'chalk and talk' would clearly not be desirable for the children. But who should change? Unfortunately, I meet comments from primary teachers about the arrogant (indeed downright bullying) attitude of some heads of mathematics – 'We are the experts who know what needs to be done'. Fortunately the scene is changing with the advent of the new General Certicate of Secondary Education. I now see practical work and groups appearing in secondary classrooms, in many cases for the first time. So perhaps a negotiated outcome is possible, but primary teachers, for all their enthusiasm, are very easily intimidated over mathematics, more than any other area of the curriculum.

In the case of governors and parents, 'seeing is believing' seems to be an effective method for a school to adopt, in bringing about conversion, conviction and support. Essentially it involves getting people to join groups of children using the new approach, or to support in some active way whatever is being attempted. If the children are seen to be enjoying and benefiting, there is no need of further argument, only general information about what is involved. Tacit acceptance of a new approach is obviously vital, but even better would be the active cooperation of parents in the learning of mathematics. How to offer help in this subject is likely to be much more puzzling and worrying to them than in the case of reading. Here, at least, encouraging children to read, listening to them and ensuring a supply of books is a more obvious means, and one where parental help has been shown to be beneficial. Several ideas have been tried by the Lakatos group, one of whose members was involved with the Mid Glamorgan Parents Project. The aims of this project were:

1 To involve parents in the development and enrichment of the mathematical language of their children.
2 To assist parents in helping to make mathematics a pleasure for their children.
3 To educate parents in the wider curriculum of mathematics.

These aims seem worth quoting as a set which might be of assistance to other authorities wishing to set up similar projects. The methods adopted by the project included a series of parents' meetings at which were demonstrated, through slides and talks, the everyday situations through which mathematical language may be introduced and developed. Booklets under the general title of *Help make maths a pleasure* were produced, accompanied by mathematical games which parents could play with their children. These games were closely related to the kinds which children would be playing in school. They could be introduced at the parents' evenings, after the talk/slide demonstrations.

Yet another approach is to involve parents in the project work of the class or school. One such project, on the theme of 'Night', produced a variety of superb displays, all linked with family efforts. The project included models of children's bedrooms, delightfully indicative of their personalities. These models give obvious leads into mathematical work (scale, plan or map drawing and so on) – mathematics *within* rather than *across* the curriculum. A possible computer microworld linked with the theme is *Suburban Fox*, a simulation in which children plan the noctural prowlings of foxes so as to survive in the town environment.

Parents are understandably anxious about their children's progress in mathematics, and are likely to see pages of sums as the best evidence of this. The use of talk and discussion in the learning of mathematics needs, therefore, to be put over to them carefully, in ways that are clear and down to earth. I note that at meetings to do with industry, or work experience, the ability to put oneself over, or to work as a member of a team, are constantly cited as desirable qualities. This constitutes a good 'external reason' for our programme. Getting children to talk about their mathematics to parents, as well as to write about it, might be helpful – maybe at meetings, where displays of work rule the roost at present. I remind you of Wiles' dictum, about providing an audience for ideas!

Parental attitudes to mathematics could have more influence on children than school and this may apply particularly to girls – 'Oh,' says mum, 'don't bother – I was never any good at mathematics either'. Enlisting parents' support and producing more informed attitudes may be a vital factor in the success of curriculum innovations. One primary school has produced its own booklet for parents, in which it encourages them to ask their children to explain things in mathematics to them. This school is one which is involved with the PrIME project I have mentioned earlier, as part of the Mid Glamorgan initiative on *Discussion-based teaching in primary mathematics* (see Ball and Brissenden, 1986). Extracts from the booklet are reproduced in the final pages of this chapter. It begins with a straightforward introduction to the ideas involved in the project, 'One of our aims in this pamphlet is to give you an insight into the work that your child will be doing and also to show you how you can help your child's mathematical learning'. It goes on to suggest that the most valuable activities are those that relate closely to the child's everyday life, and that most of these involve talking. A helpful diagram is provided (Figure 9.4).

Suggestions about activities in the home in which talk might occur are made – playing, cooking, laying the table, getting up and so on.

Talk and discussion

Developing understanding
'Tell me about the picture.'
'What do you think is going to happen?'
'What can you see?'
'Why is somebody doing that?'

Developing vocabulary
'Who is this?'
'What is this?'
'Can you see a . . .?'
'What shape is this?'
'What colour is this?'

Developing imagination
'Let's pretend . . .'

QUESTIONS THAT WILL HELP YOUR CHILD DEVELOP LANGUAGE IN MATHEMATICS

Explaining how things happen
'Why do you think that happened?'
'How did she make that?'
'I wonder how that works?'

Developing insight into how others feel
'How would you feel if . . .?'
'How do you think he might be feeling?'
'What do you think she might be thinking?'

Predicting or planning ahead
'What might happen next?'
'I wonder what they will do now?'
'What do you think is going to happen?'
'What would you do if . . .?'

FIGURE 9.4

Other ideas about conversation in the environment are also presented briefly, and the booklet concludes with the advice shown in Figure 9.5. The advice could hardly be bettered! Booklets like this could help to inform parents of developments in mathematics at their child's school, about which they might become anxious. The activities and advice in them could form a suitable theme for parents' evenings at the school. A report published by the Mathematical Association: *Sharing Mathematics with Parents* (1987) offers detailed advice about the planning of such school-based events.

Helping your child

The most useful thing you can do is to show a genuine interest in your child's work. Simply getting your child to explain how something is done mathematically will help him or her to develop understanding.

Take time to listen

Ask their opinion

Allow time for talk

Give your child opportunities to express his or her own thoughts

FIGURE 9.5

10 Mathematics and language

Language in the learning of mathematics

The main concern of this book has been with ways in which more talk and discussion can be introduced into the learning of mathematics. I justified this concern in Chapter 1 by offering four groups of reasons, of which two were to do with language - *Talk as a means of improving language skill* and *Talk as a means of developing understanding*. I suggested then that talk between teacher and pupils has to be seen as a process of *negotiation of meaning*. It is also clear that views about the nature of mathematical activity itself can have a major influence on teaching methods. Although most primary teachers are unlikely to claim to be mathematicians, the attitudes which they adopt, albeit unconsciously, will convey to pupils 'what mathematics is about'. They need, therefore, to have some appreciation of the developments that are taking place in this difficult area of the relationships between mathematics and language.

In this chapter I shall attempt to outline some of the main ideas that underlie this relationship. Readers will need to be aware that much of the research is only now developing and that there is controversy over at least one important aspect. Thus I would expect the picture in several years to be substantially clearer. There is also a lengthy time lag between the emergence of such a picture and the planning of suitable training courses and production of new primary mathematics materials. The Nuffield Scheme, for example, is based on Piagetian views which have been overtaken by fresh ideas on the role of language. While much of our material remains basically sound, I have suggested, particularly in Chapter 6, that it needs to be used in different ways.

The kinds of experience that children will meet in mathematics I summarised in my icon of Figure 1.2, which showed an interplay between doing, talking and recording. This is reproduced as Figure 10.1, but with the addition of 'Mathematical experience' as a

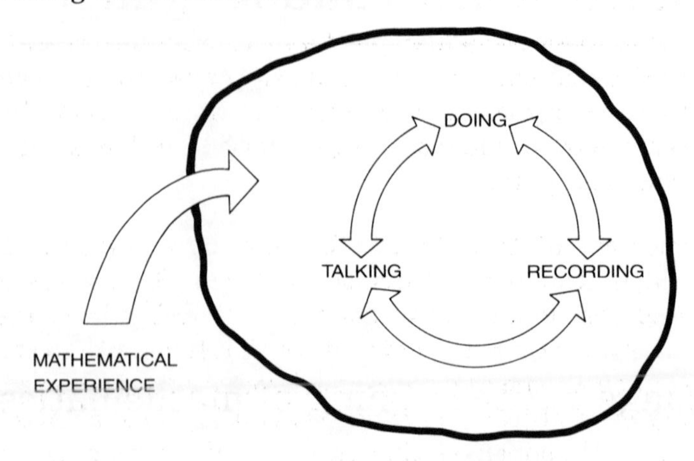

FIGURE 10.1

surround – note that I do not intend the word 'experience' to refer only to doing. The task of young children, first in the home and later at school, is to make sense of their experience. One way in which they do this, intensively researched in the 1970s, is by acquiring speech, or in Halliday's telling phrase, 'learning how to mean' in order to get things done. This natural *spoken* language is available to children when they come to school and meet the mathematical experiences of the infant classroom. Simultaneously they are learning how to read and write – a form of recording, but how does it differ from mathematical recording? We clearly regard reading and writing as distinct aspects of our teaching. Are 'mathematical speaking' and 'mathematical writing' also distinct tasks? I shall report views in the next section which show that this is likely to be the case.

The young child

Only comparatively recently have mathematics educators come to realise the importance of language in the learning of their discipline. The first time *Language and the teaching of mathematics* appeared as a theme of the four-yearly international conferences was at Berkeley, California, in 1980. Writing later, Higginson (1983) said:

> As is often the case, it seems somewhat surprising, now that we have arrived, that we didn't come sooner. Seen from close at hand the importance of the area is quite staggering, for a strong case can be made that in mathematics and language we

> have the two outstanding products of that ability which characterises our species better than any other. *Homo Sapiens* is, as Cassirer argues, more accurately *Animal Symbolicum*, the beast whose uniqueness resides largely in her ability to create and use symbols.

In an important address to that conference, reported in full in Zweng (1983), the psycholinguist Hermina Sinclair described various findings about young children's thinking in connection with alphabetic writing. She pointed out that humanity took a long time to work out such systems, which involve a degree of reflection on language as an object of knowledge in itself. Our system is based on an idea - the phoneme - that has no real existence in speech, where the only more or less natural unit is the syllable. Like most symbol systems, alphabetic writing does not represent all the properties of that which it symbolises - intonation, speed and so on. It also has properties of its own - blanks between words, left-to-right and top-to-bottom scanning, punctuation, which have no correspondents in speech. Young children are surrounded by examples of this system, and from the age of three or four single out letters and numbers from their environment, distinguishing them from other squiggles such as decorations, wallpaper patterns and so on.

By the age of four, many children make the important step of attributing meaning to these squiggles, and form their own theories about it. Very often these are quantitative in character - number or size give meaning to letters, for example a picture of three dogs needs three squiggles underneath. The meaning of letters may also be linked with the object they appear on, or the accompanying picture - it is easy to see how this happens. But there is no link with the sounds the letters represent for us. When there is no picture or object, letters are thought to represent the intentions of the person who wrote them, so that only he can read them. Thus four year olds were found to believe that the postman wrote the addresses on envelopes so as to know where to bring them! The first squiggle/sound link established is a syllabic one - a highly plausible theory, since syllables are a natural unit of speech production. All these theories come from the children, of course, not from adults or older children. There is nothing natural about the alphabetic system, Sinclair concludes, and those given the task of teaching it cannot assume that it is simply a matter of learning the phoneme-grapheme correspondence. Nor can we assume that the children come as 'empty vessels' with no prior meanings which they attribute to the system.

Hawkins (*op cit*) points out that English spelling does indicate pronunciation but gives lexical and syntactical information besides. Thus the '-ed' in final position is a marker of past tense, as in 'walked', not a signal of the actual sound, 'walkt'. English spelling works very well for those who have a good knowledge of the spoken language, he says, but makes it hard for a beginner who does not know the language well. A variety of sub-systems operate, so that, for example, Bernard Shaw's notorious suggestion that 'fish' might be spelt 'ghoti' is based on a misunderstanding, since the symbol 'gh' cannot represent the sound 'f' in *initial* position in any word spelt in English. The language underlying the spelling is adult language, and young children need adult time and dialogue in order to learn to read.

Sinclair went on to question the corresponding assumptions made about the teaching of arithmetic to young children. Just as with written language, young children have theories about written numerals and equations, which may not fit the systems (such as place value) which have taken centuries for humanity to develop. Moreover, the system they are learning for reading and writing is unhelpful to, perhaps in conflict with, the mathematical system. Most importantly, Sinclair questions the assumption that developing an understanding of arithmetical operations is the same as writing them in symbolic forms. The recording of mathematical ideas is a separate teaching task which presents difficulties of its own, analogous to those presented to children in learning to read and write.

At the time of Sinclair's address, research into the mathematical aspects she describes was very limited. Since then, however, the work of Hughes (1986) has thrown considerable light on young children's thinking about number. He has shown how they can make effective use of their own iconic recordings – readers may well have seen the television recording of his experiment in which a young child keeps track of the contents of three tins using his own symbols on the lids. Hughes thinks that mathematical games could be helpful to children in developing understanding. The following extract from his book highlights some of the strange features of mathematical language which young children have to acquire in order to learn how to mean in a new way.

Martin Hughes: What is three and one more?
Ram (the child): One more what?
Martin Hughes: Just one more, you know?
Ram (disgruntled): I *don't* know. (*op cit*, p 45)

In mathematical talk, 'one', 'two', 'three', etc are detached from any referent, but used as if they were things – 'mathematical objects of thought'. Ram has presumably not yet learnt to use language in this new way, hence his annoyance with the experimenter. His progress will depend as much on advances in his use of language as on experience with objects such as counters. I will develop this idea further in the next section, concluding this one by listing the language teaching tasks that have been identified.

Natural language

- continue to develop natural speech
- teach how to read natural language
- teach how to write natural language

Mathematical language

- teach how to speak mathematically
- teach how to record mathematics
- teach how to interpret mathematical recordings

Both these forms of language are involved in the main task, that of developing understanding of mathematical concepts. An interesting and possibly significant controversy has arisen over the respective roles played by the three aspects of doing, talking, and recording (as shown in Figure 10.1 above.) Two of these aspects are obviously concerned with language. Which is 'in the driving seat', so to speak, in the learning of mathematics, the language or the doing? Conventional wisdom, embodied in most current mathematics schemes, is that 'experience' (interpreted as practical activity with objects) is the essential basis – as exemplified by Nuffield's 'I do and I understand'. But is not language a part of children's experience? Is it possible that it is the part that really matters?

Language as experience

Before describing the details of this controversy, I need to make it clear that no-one is arguing that practical activity is not essential on many occasions. What has to be clarified is the exact role that it is playing. At other points in children's learning, teachers will need to decide whether what is needed is more practical work or more work involving language. In the latter case children will discuss and reflect on mathematical language in itself, as suggested by Skemp's functionings in Chapter 6. The controversy arose and developed in the journal *Mathematics Teaching* during 1985–6, beginning with an article by Tahta (No 112) 'On Notation'. Tahta drew attention to the research of Corran and Walkerdine (1981), using it to cast doubt on current orthodoxies such as 'that action and experience necessarily precede thinking' or that 'concepts develop independently of language'. Reasoning demands reflection on what Tahta calls *metonymic relations of language*, that is, relations between signs without reference to what they signify. He also introduces the linguistic term *metaphoric relation* into the discussion, meaning a movement from the sign to what it is that is signified. Such a movement would occur in mathematics whenever a particular referent, such as counters, rods, blocks, etc, is introduced into an activity. Examples of discussions carried on at a metonymic level are (I think) the accounts 'What does "and" mean?' and 'Trouble with decimal places' in Chapter 3. I shall return to the latter shortly.

Liebeck responded to Tahta's article in issue No 114, 'In Defence of Experience'. She advanced a four-part, or ELPS, model for the learning of mathematics: E(xperience with familiar objects), (spoken) L(anguage), P(ictures) and S(ymbols). This she illustrated with examples drawn from different mathematics schemes. The model appears to suggest that these aspects are disjoint and chronologically ordered, with 'experience' as the essential foundation. The controversy continued with a further article by Pimm (No 116) called 'Beyond Reference'. Pimm criticised Liebeck for apparently separating language from experience (something I have been careful to avoid in Figure 10.1) arguing that 'With mathematics it is important to perceive language as experience as well as mediator between experience and the individual.' He also criticised her apparent concentration on written mathematics rather than talk. Pimm believes that we need to consider the range of functions of language in mathematics, rather than mere quantity, whether it be of talk or recording. With this point I strongly agree – the suggestions in this book are directed not only at changing the balance of classroom talk, but also, I hope, at altering its nature and quality. Pimm's major point is probably contained in the following paragraph:

> . . . children have to acquire the sense in which a new linguistic construction is being employed (use of number words as nouns); that is, they have to learn how to mean as a mathematician does. In particular, they will not necessarily do so by further experience with manipulation of, counting of or discussion of particular sets of objects, as this only involves the other, adjectival use of number words. This (possibly dubious) use of equipment in an attempt to assist in the learning of a *linguistic* phenomenon will recur when we come to look at place-value.

Pimm was referring to the account reproduced earlier, in which Ram got annoyed with Hughes about 'three and one more'. But in several of the accounts in Chapter 3, we see children who have made this important linguistic step. Take '*Unifix* steps', for example, where Leanne counts up to six aloud while pointing to her rods (Figure 3.5). She can apply this vital linguistic achievement to constructing her set correctly, although not yet to the more difficult task of step building. Note Matthew's instructions to Kimberley about this, 'You need one and then two and then three and then four and then five . . .'. In a later account, 'In the touching corner', Dale 'counted as far as he could' in describing the toy centipede, and the teacher expresses concern about Tanusuree's counting and finger matching. In contrast, in Chapter 6, Vicki gets lost in making 'two add two add two add two . . . add two . . .' with her twenty *Unifix* cubes – she has not thought yet about counting her sets of two. It could be argued that learning to use an agreed sequence of sounds – whether from the teacher or from other children – is the vital ingredient for success in each case.

Much of the controversy centred around another important concept, place value. Here Corran and Walkerdine argued, with Pimm's clear approval, that it is the teacher's language, as she shows the children how to bundle sticks into tens, tens of tens etc, which provides the cues from which the children construct a system of place value. Acquiring this grouping convention – a property of language – then enables them to organise their practical activities. This view effectively reverses the conventional formulation, that they are 'abstracting place value from a series of experiences'.

The account of Ann's 'Trouble with decimal places' is interesting in the light of this. The discussion is almost entirely concerned with mathematical language, and centres around the development of a 'ten times' idea to produce a self-consistent schema. Notice how the spoken forms of number are helpful to Ann at first – 'Four hundred and fifty-seven' helps her to recover the hundreds, tens and units – but then there is no such help when she moves to decimal forms. 'Twenty-seven point five six eight' offers no clues about the nature of the numerals on the right of the decimal point. How should we speak decimals? One sometimes hears barbarisms such as 'twenty point fifty-two' which I think should be avoided. The extent to which practical activities are to do with conventions is revealed in Figure 3.7. The block which in previous activities was regarded as 1000 units (when units were 1 cm cubes) is now treated as 'one whole'.

Metonymic moves – 'playing around with language' – will not necessarily lead to correct results, of course, and testing has to be carried out in several different ways, via Skemp's functionings. The account of 'What does "and" mean?' involves a great deal of this, and shows a group reflecting on and discussing the variety of mathematical statements they generate themselves. It looks as if mathematical situations based around *Trio Tricks* might be a good way of provoking this sort of activity. Mathematical thinking is often rather like this metonymic play, and long periods may elapse before ideas are accepted. For example, negative values were at one time rejected as 'false roots' to equations, and as late as 1800 mathematicians were debating what the result of -1×-1 should be. Mathematical language often provoked invention, however, a good example being the way in which power notation was gradually extended. To show this in a much shortened form, suppose I construct a table of powers of 2, using results like $2^2 = 2 \times 2 = 4$, $2^3 = 2 \times 2 \times 2 = 8$ etc. I get

Power of 2	1	2	3	4	5	6
Value	2	4	8	16	32	64

Now I have a table of pairs of values, so I can plot these as points on a graph, as in Figure 10.2. The power of two is read horizontally, the value vertically. Once the points are plotted, it is tempting to join them up with a nice smooth curve – a graph of powers of two. Suddenly there is something new, because I can read off from the horizontal axis values which did not exist before I drew the graph! For instance, '2 to the power of 2.5' (whatever that might mean) appears to be equal to 5.6 approximately, as shown in Figure 10.2. Trying to fit this new mathematical object into the existing scheme of things is now the logical task that has to be carried out. Another new mathematical object created is 2°, whose meaning also has to be discussed.

One of the properties of 'whole number powers' is the 'law of addition'. For example, since

$$4 \times 8 = 32,$$

we observe that

$$2^2 \times 2^3 = 2^5 = 2^{2+3}.$$

From relations like this we conclude that $2^m \times 2^n = 2^{m+n}$, as long as m, n are whole numbers. Is the new form consistent with this law of addition of whole number powers? If it is, then we should have, for instance,

$$2^2 \times 2^{2.5} = 2^{2+2.5} = 2^{4.5}.$$

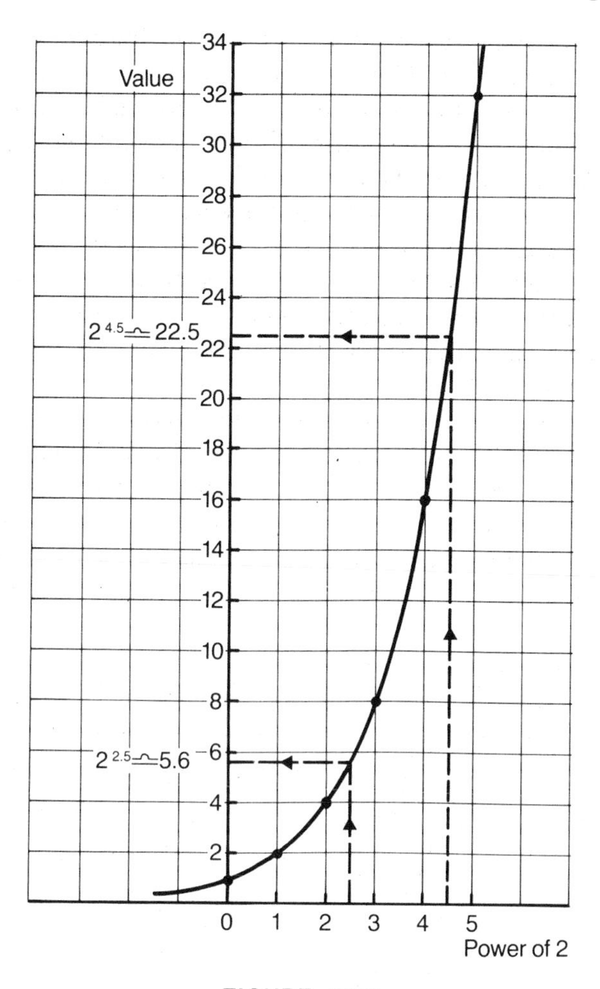

FIGURE 10.2

This can be tested by reading off values from the graph in Figure 10.2. We know that $2^2 = 4$ and $2^{2.5} \simeq 5.6$, so $4 \times 5.6 = 22.4$. Is this the same as the value for $2^{4.5}$ read from the graph? Reading from the horizontal value at 4.5 gives a value of 22.5 which agrees with the previous one approximately. So the new kind of power discovered from our graph seems likely to be consistent with previous mathematics. I will leave readers to try to find a consistent meaning for 2°, if they have not met it already. This short account illustrates how a metonymic move still has to be tested against other mathematics. The choice of a result for -1×-1 (Is it to be $+1$ or -1?) involves a similar search for consistency of thought in the resulting structure. This was one of Skemp's functionings described in Chapter 6.

I have gone into some detail over this controversy about the relationship between language and practical activity in the learning of mathematics, because it could have an important bearing on our teaching methods, now that the importance of language is recognised. Some clear thinking may be needed, when we plan our activities, on the part played by doing, on the one hand, and by talking and recording, on the other. I am quite sure, in agreement with Pimm, that the mathematical experience of children includes all three. But the interplay between them is likely to be untidy, rather than some neat sequence – as suggested by the icon I have chosen in Figure 10.1. I cannot settle the controversy, of course, but since the publication of the articles mentioned, Pimm has written in much more detail about 'speaking mathematics' (Pimm, 1987). As for mathematics and the real world, he has this to say:

> Far from mathematics being all around us, I offer the alternative tenet that mathematics is *only* inside us, as language is, and the signs and symbols, whether of natural language or mathematical language, are projections of it into and onto the world. Since mathematics is not of the material world nor has very much to do with it, an activity-based approach relying solely on physical objects is surely limiting and inappropriate.

I believe that teachers would do best to think of mathematics as a way of knowing, about ourselves and about the world – a creative activity, carried on by people, and involving meaning. A contrasting view is that of mathematics as a body of knowledge, which pupils accept uncritically. The way of knowing outlook seems in general accord with the philosophy of Lakatos, which I described briefly in Chapter 1. He, you will recall, was opposed to the formalist view of mathematics, which sees it as a kind of elaborate system of rules which dictate how 'the game is to be played'. People, and perhaps meaning, seem to have disappeared from this picture. Lakatos thought of mathematics as a creative activity carried on by arguments about proofs and refutations. The mathematicians Davis and Hersh, in a widely-regarded book, *The Mathematical Experience* (1981, p 358), speak out in strong support of the Lakatos view. However, it would be inappropriate to delve deeply here into these questions of mathematical philosophy, let alone how we seem to be able to use mathematics to explain, control or predict events in the world!

Reading and writing mathematics

I want to turn now to a brief discussion of the tasks highlighted by Sinclair, the reading and writing of mathematics. She said, you recall, that writing mathematics had its own difficulties, distinct from those of writing natural language. Moreover, it was not synonymous with gaining an understanding of mathematical concepts. I begin by describing a distinction made by Woodrow (1982) between signs and symbols. Adults have little difficulty in handling signs in everyday life. Figure 10.3 shows several examples, which readers will respond to immediately. Traffic signs, laundry signs, signs on cars or electronic equipment are all around us and are used ever more widely in the modern world. All these examples have the distinct features of a sign – they represent a single low level naming concept which identifies a single, non-adaptable idea. They have very little redundancy – that is, their meaning is seriously affected if they are altered markedly; they are not interdependent and their meanings are unaffected by neighbouring signs. The last point is obvious from Figure 10.3!

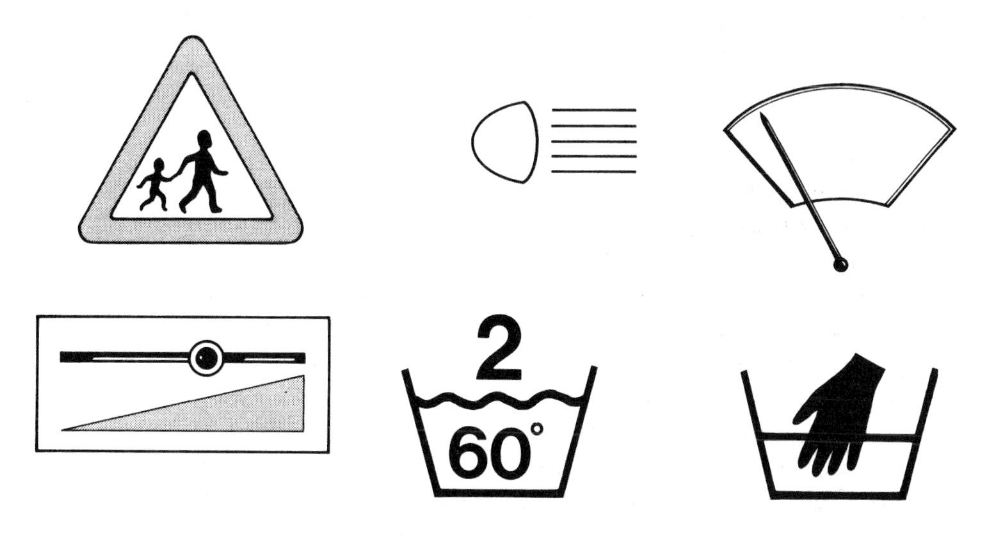

FIGURE 10.3

Compare this with the situation in written language based on an alphabetic system like English. Here there is plenty of evidence for considerable redundancy. We are able to make sense of passages even when there are misspellings and letters or even words and phrases ommitted. The well-known experiment of covering up the top half of a line of print shows that most people are still able to read using the cues that are left. In many techniques used in reading we apparently scan the letters and words only partially. No individual letter has any meaning in itself, and thus there is great interdependence between them in conveying any meaning. Somehow, we build meanings from the text. Children will meet plenty of this natural written language around their mathematical symbols, of course, in workcards and textbooks. But when they try to extract meaning from mathematics recordings, the constraints are rather different.

Consider, for example, the meanings which attach to the symbol '2' in each of the following:

$$212,\ \tfrac{1}{2},\ 2,\ 2^2,\ 2\text{ o'clock},\ (2,3)$$

Notice that there is little or no redundancy, certainly at school level. Whereas in reading the individual symbols contain no meaning, in mathematics the meaning of each symbol is vital. Moreover, even though each symbol has this essential meaning, that meaning is affected by neighbouring symbols, requiring the learner to 'hook on' to different schemas in subtle ways. The left to right scanning habit painstakingly acquired in ordinary reading can conflict with what is needed in

$$3 + 4 \times 5$$

or

$$3 + 2x = 7 + x$$

What seems to be required here is a kind of 'spatial scanning', which holds an overview of the entire statement, rather like studying a diagram. Skemp has pointed out that there are few spatial relationships available to convey the many varieties of mathematical meaning – little more than up-down, left-right, combined with the use of brackets or superscripts and subscripts. His example is our old friend place value. When Ann looks at '457' for instance, there are several layers of meaning for her to unravel. At the first level, each symbol has its own distinct meaning. At the second level, the left to right order of the digits corresponds to three powers of ten (or 'place values'). Next the digits are understood to be multiplied by these powers of ten, to give 'four hundreds', 'five tens' and so on, and finally the addition of these results is the actual number meant, so that

$$457 \text{ means } 4 \times 10 \times 10 + 5 \times 10 + 7$$

Only the first level of meaning is explicitly represented by different symbols, the second is implied by the left to right order, while the third and fourth have to be inferred from the fact that there is more than one digit in the numeral.

The result of this shortage of spatial relationships is that similar forms are pressed into service to symbolise markedly different meanings, presenting obvious difficulties to the young learner. It is easy to think of examples, such as the varied use of the + sign, first in adding whole numbers, then fractions, then (possibly) vectors, or juxtaposition, which can mean place value with numbers but multiplication in algebra. Or consider the meanings that mathematicians attach to the brackets in statements like

$$3(a + b) = 3a + 3b, y = f(x)$$

The premature introduction of symbols to represent mathematical structures can lead to pupils developing inflexible schemas of meaning, which they cannot later adapt to different situations. Woodrow criticises two teaching approaches in this respect. The traditional model of examples followed by practice leads to an uncritical acceptance of the teacher's schema by most pupils, which cannot later be adapted except by the more gifted. Individualised schemes, which as we have seen are typical of many primary classrooms, do build a broad base of low-level concepts in which symbols are related to limited content in short pieces of work. But the higher level relationships may not be developed from these numerous individual tasks, whose effects are quickly dissipated and forgotten. The meanings attached to the symbols are thus as narrow and lacking in adaptability as before, although for different reasons.

Woodrow argues that we need, in future mathematics schemes, to adapt mathematical symbolism to the learner and follow a careful and structured plan in order to teach our pupils how to read mathematics. Skemp's suggestions include the sequencing of new material so that it can always be assimilated to a conceptual structure, not just memorised in terms of symbolic manipulations which lack meaning. This is a considerable challenge to the writers of schemes, as well as to teachers, of course. The temptation will always be to fall back on 'This is how you do it', as we saw in Chapter 1. Skemp considers that it helps if we stay longer with spoken language, whose connections with thought are intially much stronger than those between thought and written words or symbols. Finally, as was suggested in Chapter 6, he advocates the use of informal, 'transitional notations' as bridges to the formal, highly condensed notations of traditional mathematics. This activity, which involves inventiveness on children's part to enable them to convey meaning successfully to their colleagues, may help them to be more flexible in adjusting to changes in meaning of mathematical symbols as they progress.

Quadling (*op cit*) summarises the findings of the Assessment of Performance Unit on these aspects of reading and writing mathematics. Chapter 3 discusses 'Number, Language and Context' while chapter 4 looks at 'Diagrams' as a means of communication, and the ways in which they can both help and hinder. Another helpful reference, edited by Shuard and Rothery (1984), discusses the ways in which children read mathematics texts. Their approach was criticised by Wing (1985), as being too prescriptive in outlook, in stating that in mathematics, reading is 'getting the meaning from the page'. What might be the consequences, he argues, of seeing reading as an activity in which a reader puts meaning on to a page? Such a view sees meaning as the creation of the reader, not something which has an independent life of its own. There certainly seems to be a place for written learning materials which encourage children to create mathematics, alongside the need to read 'accurate' meanings from text, as in word problems. Wing points to the way in which text forces the reader to focus on internal, metonymic relations rather than external, metaphoric meanings. This could be helpful for the development of mathematics, but the absence of a speaker endows the text with a kind of authority which runs counter to the active role of the learner in the negotiation of meaning. Over-reliance on a print-based system of learning mathematics could thus lead to a very narrow, passive outlook, unless materials of the type Wing describes were included.

Support for the idea that the reader plays an active role in constructing meaning comes from Davis (1984), in an important, but difficult, reference. Davis quotes a paragraph (p 45) with further 'comprehension' questions – very typical-looking material until he makes it clear that the replies to the questions, which are easily supplied by normal adults, are *not* actually contained in the paragraph! The information could only come from readers' memories, triggered somehow by the sentences in the paragraph. Davis quotes the research of Minsky, Schank and others, which disputes the view that reading is a process of careful extraction of information contained on a printed page. Minsky's theory involves the cuing by the text of certain 'frames' from the reader's memory, in which various 'values' are inserted. According to this theory, different readers reading the same text would extract different information, because the frames in their memories would differ. Large sections of Davis's book are concerned with the search for, and description of, various common 'frames' of mathematics which research has revealed in learners. Davis uses 'frame' in a generally similar manner to the term 'schema' which I have adopted in this book.

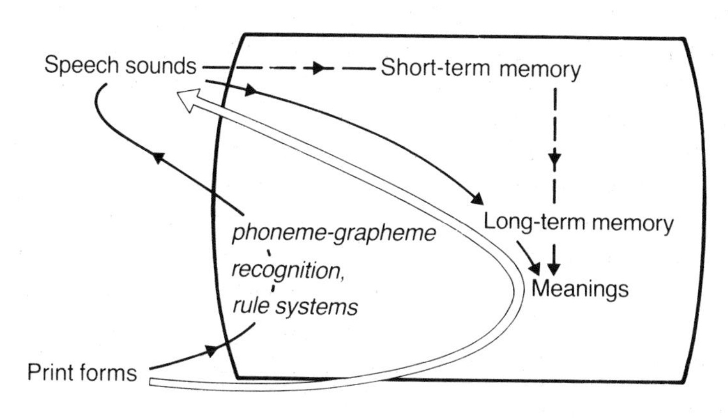

FIGURE 10.4

The teaching of reading is still capable of provoking considerable controversy. During 1987, for example, there was much debate over ways of helping children with special needs (such as Down's syndrome children) in their language development. One school of thought argued that print could be useful in this process, and might take precedence over speech. For all normal children, however, it seems accepted that spoken language is the foundation on which reading and writing are built, in a manner suggested in Figure 10.4. The speech sounds fade rapidly from the short term memory, what is transferred to the long term memory is the meanings of speech. Print forms are understood partly by linking with these already understood meanings (and so indirectly with speech), or by using the various sub-systems described by Hawkins (*op cit*) and partly by recognising phoneme-grapheme correspondences. A nonsense collection of letters like PLAX is more easily remembered than one like ZBQT because it consists of letter associations which occur in English spelling, so that it can be pronounced to oneself.

Constructing a similar diagram for mathematics looks difficult! For one thing, as we move away from number words and elementary mathematics, the link with speech forms fades and disappears, leaving mainly (or only) a link between symbols and meanings, or rules for manipulating the symbols. (Recall how Ann had whole number words to say, but found it more difficult to say decimal numbers.) Talking mathematics at more advanced levels seems to consist of relating concepts or meanings with others, announcing intentions or explaining complex operations that are being performed with symbols – the metonymic moves mentioned

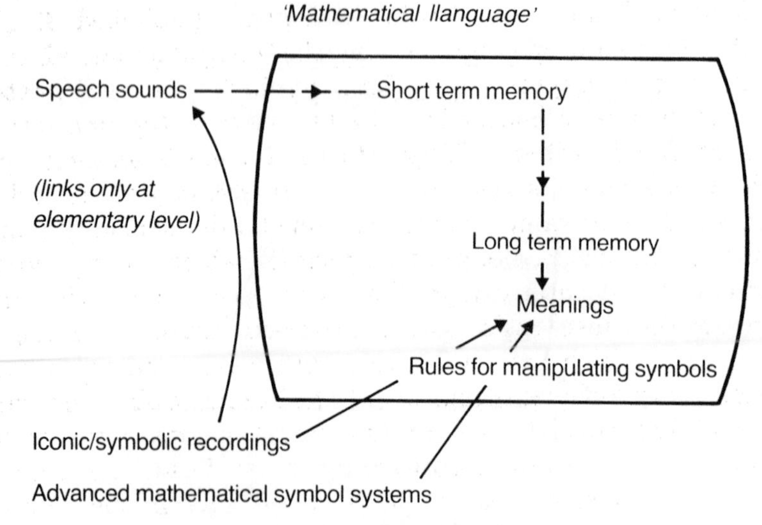

FIGURE 10.5

earlier. Most of the power of advanced mathematics resides in such manipulations of symbolic recordings, in which whole chunks of mathematics are treated as if they were a single 'object'. Nevertheless, mathematicians manage to lecture on their subject at universities, so speech must have this mainly explanatory role. Our concern is of course with children, where there is no doubt that speech, as well as recording, has a vital part to play in helping them to gain understanding. But there are not likely to be the direct links between sounds and recordings that exist in natural language. For such reasons, Skemp has suggested that the use of the term 'language' in connection with mathematics is misleading. He prefers to describe both the spoken and written forms as a 'symbol system', a view with which I agree. During a long discussion on the matter at Gregynog, in mid-Wales, Skemp coined the term 'llanguage' to distinguish mathematics from natural language. If the arguments put forward in this section of the chapter are accepted, then the teaching of 'mathematical llanguage' is an important task, to be distinguished from the development of language, whether in speech or in written forms. In Figure 10.5 I have attempted to construct a diagram for mathematical llanguage, corresponding to Figure 10.4 for natural language.

Language in problem solving

How does language function during problem solving, and in particular during the investigative kinds of activities which were described in Chapter 6? Goldin, in a paper published in 1982, described four kinds of language which he thought appeared. I have already referred to some of their functions in discussing the extract about 'Leapfrogs' (Chapter 3) and later the investigation into the 'Tower of Hanoi' problem (Chapter 6). First, there is the language in which the problem is originally posed – the language of the problem statement. In the examples just mentioned, this consists of stating the rules of the puzzle, along with a question which sets up a goal to be achieved – to interchange the frogs, or transfer the discs, according to those rules. Other problems involve 'setting up information', which may be quite complicated. Linked with this function of problem-posing is what Goldin calls 'non-verbal language'. This consists of diagrams, graphs, charts, some of which might be offered at the start, others generated during the process of solution. Examples of both of these were offered in the 'Tower of Hanoi' account. Often the first step in the solution process is to investigate them, and to translate the problem statement into such forms. One important function seems to be to clarify the meaning of the problem, to gain insight by studying instances. The third form described by Goldin is 'notational language', which is often suggested by the non-verbal forms of language. Developing a helpful notation is very often the key to solution – you can see how this was so with the tabulation in the 'Leapfrogs' and the notation I invented for the 'Tower of Hanoi'. They help us to see patterns more clearly, thus generating conjectures about the problems and suggesting lines of attack on it.

All three forms of language discussed so far involve working directly in the problem, so to speak, using either non-verbal or notational language. But there is a fourth type of language, elicited by these three, which Goldin calls 'planning language'. This language is used to talk about the other three kinds, to keep track of

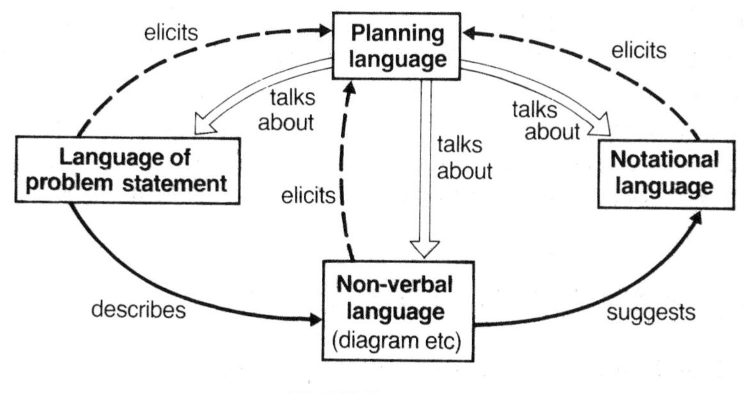

FIGURE 10.6

attempts, to review and make plans about what to try. Sometimes it is the teacher who is effectively the planning leader – this is probably the case in the 'Leapfrogs' account. With children's groups the planning language may be sporadic or ineffective. The strong recommendation arising from studies of problem solving is that we need to make groups more aware of this function, and offer guidance to them about planning. The mathematician George Polya is the name most prominent in this movement, but much other work on the same lines is described in Shuard, (1986a). Thus teachers might at first tread a line between offering guidance and holding back, but later insist that groups develop their own planning. The relationships between Goldin's four types of language are shown in my final diagram, Figure 10.6.

Reviewing the nature of the task

I began this book by describing, in the first two chapters, the 'Nature of the task'. What followed, in Parts Two and Three was a detailed programme which suggested how primary schools might respond to the challenge to review, and probably modify drastically, the way in which they taught mathematics to children. In these final chapters I have returned, in a sense, to outlining further possible tasks, by setting mathematics within the wider context of language development. It should be clear to readers that there has been a major change in outlook about the learning of mathematics, with the realisation of the importance of spoken language in the development of meaning. The rapid developments in technology and artificial intelligence add a further exciting dimension to the challenge. Just as interesting to the mathematician is the manner in which the ideas relate to philosophical views about the nature of mathematics itself. I hope that the programme put forward here will be helpful to primary teachers in developing their own thoughts and methods so that their children can 'learn how to mean as mathematicians do'.

Bibliography

Association of Teachers of Mathematics/Mathematical Association (1986) *Calculators in the Primary School* Leicester, The Mathematical Association.

Austin, J.L. and Howson, A.G. (1979) 'Language and Mathematical Education' in *Educational Studies in Mathematics, 10*, 161–197.

Ball, G. (1986), 'The Effective Use of the Computer in the Infant Classroom' in *Micro-Scope, 18, 9–12.*

Ball, G. and Brissenden, T.H.F. (1986) *PrIME in Mid Glamorgan* Mid Glamorgan Education Authority.

Barnes, D. and Todd, F. (1977) *Communication and Learning in Small Groups* London, Routledge and Kegan Paul.

Boydell, D. (1981) 'Classroom organisation' in Simon, B. and Willcocks, J., *Research and Practice in the Primary Classroom*, London, Routledge and Kegan Paul.

Brighouse, A. *et al* (1983) *Directions 1, Left and Right* Thomas Nelson and Sons Ltd.

Brissenden, T.H.F. (1980) *Mathematics Teaching, Theory in Practice* London, Harper and Row.

Brissenden, T.H.F. (1984) *Implications of the Cockcroft Report for Secondary School Mathematics* Swansea, Swansea University College Faculty of Education.

Bruner, J.S. (1973) *Beyond the Information Given: Studies in the Psychology of Knowing* New York, Norton.

Buck, S. (1986) 'ISLAM – but is it art? An Investigation into Patterns from Digital Root Sequences' in *The Times Educational Supplement* 9.5.86.

Bullock, A. (Lord Bullock) (1975) *A Language for Life (The Bullock Report)* London, Her Majesty's Stationery Office.

Burton, L. (Ed) (1986) *Girls Into Maths Can Go* London, Holt, Rinehart and Winston.

Button, L.B. (1981) *Group Tutoring for the Form Tutor* London, Hodder and Stoughton.

Cockcroft, W.H. *et al* (1982) *Mathematics Counts (The Cockcroft Report)* London, Her Majesty's Stationery Office.

Corran, G. and Walkerdine, V. (1981) *The Practice of Reason, Vols. 1 & 2* Institute of Education, London University.

Davis, P.J. and Hersh, R. (1981) *The Mathematical Experience* Boston, Houghton Mifflin.

Davis, R.B. (1984) *Learning Mathematics* Beckenham, Croom Helm.

Dean, J. (1983) *Organising Learning in the Primary Classroom* London, Croom Helm.

Department of Education and Science (1984) *English from 5 to 16* London, Her Majesty's Stationery Office.

Department of Education and Science (1985) *Mathematics from 5 to 16* London, Her Majesty's Stationery Office.

Dienes, Z.P. (1971) *Modern Mathematics for Young Children* Harlow, Educational Supply Association.

Easley, J. (1984) 'A Teacher Educator's Perspective on Students' and Teachers' Schemes: Or Teaching by Listening' paper presented at the Conference on Thinking, Harvard Graduate School of Education, 1984.

Fitzgerald, A. (1985) *New Technology and Mathematics in Employment* Birmingham, Department of Curriculum Studies, University of Birmingham.

Floyd, A. (1981) *Developing Mathematical Thinking* London, Addison-Wesley for the Open University.

Galton, M., and Simon, B. (1981) 'ORACLE: its implications for teacher training' in Simon, B. and Willcocks, J. (Eds) *Research and Practice in the Primary Classroom* London, Routledge and Kegan Paul.

Goldin, G.A. (1982) 'Mathematical Language and Problem Solving' in *Visible Language Vol XVI No 3* 221–238.

Gray, D. (1979) *Flow and Control in Mathematics 5–13 (West Glamorgan Guidelines)* Basingstoke, Globe Education.

Halliday, M.A.K. (1973) *Explorations in the Functions of Language* London, Edward Arnold.

Halliday, M.A.K. (1975) *Learning How to Mean* London, Edward Arnold.

Hames, J. (1982) *Make Thinking Visible* published in conjunction with Gridsheets by Excitement in Learning, 88 Mint St. London.

Hawkins, E. (1987, 2nd edition) *Awareness of Language* Cambridge, Cambridge University Press.

Her Majesty's Inspectorate (1986) *Taking Stock: Primary Schools* Cardiff, The Welsh Office.

Higginson, W. (1983) 'Threeks, Rainbrellas and Stunks, Reaction to Hermina Sinclair's Plenary Lecture,' in Zweng, M *et al Proceedings of the Fourth International Congress on Mathematical Education* Birkhauser, Boston.

Hislam, N. (1985) 'Woodlice and Infants' in *Every Child's Language, An In-Service Pack for Primary Teachers, Book II* Clevedon, Multilingual Matters in association with the Open University.

Hughes, M. (1986) *Children and Number* Oxford, Basil Blackwell.

James, N. and Mason, J.H. (1982) 'Towards Recording' in *Visible Language Vol XVI No 3* 239–258.

Jacques, D. (1984) *Learning in Groups.* Beckenham, Croom Helm.

Jones, P. (1986) 'Groby Oracy Project' in *The Times Educational Supplement* 3.10.86.

Kerslake, D. *et al* (1982) *Mathematics, Part 6 of Language, Teaching and Learning* London, Ward Lock Educational.

Kinder, R.F. (1969) 'Learning to Listen' paper presented at 6th Annual Conference of United Kingdom Reading Association (UKRA).

Lakatos, I. (1976) *Proofs and Refutations* Cambridge, Cambridge University Press.

Lea, D. (1986), 'New Maths Horizon' in *The Times Eductional Supplement* 16.5.86.

Liebeck, P. (1986), 'In Defence of Experience' in *Mathematics Teaching* 114 36–38.

Mason, J.H. *et al* (1982) *Thinking Mathematically* London, Addison Wesley.

Mathematical Association (1987) *Sharing Mathematics with Parents* Leicester, The Mathematical Association.

McMahon, A., *et al* (1984) *Guidelines for Review and Internal Development in Schools: Primary School Handbook* York, Longmans Resources Unit.

Mercer, N. (1985), 'Communication in the Classroom' in *Every Child's Language, An In-Service Pack for Primary Teachers, Book I*, Clevedon, Multilingual Matters in association with the Open University.

Mottershead, L. (1978) *Sources of Mathematical Discovery* Oxford, Blackwell.

Mottershead L. (1985) *Investigations in Mathematics* Oxford, Blackwell.

Nash, R. (1973) *Classrooms Observed* London, Routledge and Kegan Paul.

Nash, R. (1976) *Teacher Expectations and Pupil Learning* London, Routledge and Kegan Paul.

Needham, J. (1985) 'Activities to Develop Co-operation and Communication between Pupils' in Mercer, N. *Every Child's Language* Book II, Clevedon, Multilingual Matters in association with the Open University.

Open University *Developing Mathematical Thinking (EM 235) Topics 3 and 5*, Milton Keynes, The Open University.

Pearson, S. (1986) *Primary Education: the Role of the Microcomputer* Cardiff, MEU Cymru.

Pearson, S. (1985) *Walk* programme developed for the MEP, Tecmedia Productions.

Phillips, T. (1985) 'Discourse Development after the Age of Nine' in Wells, G. and Nicholls, J. (Eds) *Language and Learning - An Interactional Perspective* London, Falmer Press.

Pimm, D. (1986) 'Beyond Reference' in *Mathematics Teaching 116*, 48–51.

Pimm, D. (1987) *Speaking Mathematics* London, Routledge and Kegan Paul.

Quadling, D.A. *et al* (1985) *New Perspectives on the Mathematics Curriculum* London, Her Majesty's Stationery Office.

Schools Council (1981) *The Practical Curriculum (Schools Council*

Working Paper 70) London, Methuen Educational, for the Schools Council.

School Mathematics Project (1984) *Pointers: Planning Mathematics for Infants* Cambridge, Cambridge University Press.

Shuard, H. and Rothery, A. (Eds) (1984) *Children Reading Mathematics* London, Murray.

Shuard, H. (1986a) *Primary Mathematics Today and Tomorrow* York, Longmans.

Shuard, H. (1986b) *Primary Mathematics Towards 2000* Presidential Address to the Mathematical Association, Leicester, The Mathematical Association.

Skemp, R.R. (1976) 'Relational Understanding and Instrumental Understanding' in *Mathematics Teaching 77*, 20–26.

Skemp, R.R. (1982) 'Communicating Mathematics: Surface Structures and Deep Structures' in *Visible Language Vol XVI, No 3*, 281–288.

Skemp, R.R. (1986) *Primary Mathematics Project for the Intelligent Learning of Mathematics* Department of Education, University of Warwick.

Straker, A. (1984) *Picfile* Cambridge Micro Software, Cambridge University Press.

Tahta, D.G. (1985) 'On Notation' in *Mathematics Teaching 112*, 49–51.

Tann, S. (1981) 'Grouping and Group work' in Simon, B., and Willcocks, J. *Research and Practice in the Primary Classroom* London, Routledge and Kegan Paul.

Tough, J. (1976) *Listening to Children Talking: a Guide to the Appraisal of Children's Use of Language* London, Ward Lock Education in association with Drake Educational Association.

Tough, J. (1977) *Development of Meaning* London, Allen and Unwin.

Vygotsky, L.S. (1971, 6th edition) *Thought and Language* Cambridge, Mass., M.I.T. Press.

Wells, G. (1984) 'Talking with Children: the Complementary Roles of Parents and Teachers' in Donaldson, M. *et al* (Eds) *Early Childhood Development and Education* Oxford, Basil Blackwell.

Wiles, S. (1981) Programme Notes for the BBC Schools Series *Talkabout* London, BBC Publications.

Wiles, S. (1985) 'Language in the Multi-Ethnic Classroom' in Wells, G. and Nicholls, J. (Eds) *Language and Learning - An Interactional Perspective* London, Falmer Press.

Wing, T. (1985) 'Reading Children Reading Mathematics' in *Mathematics Teaching, 111*, 62–63.

Woodrow, D. (1982) 'Mathematical Symbolism' in *Visible Language, Vol XVI, No 3*, 289–302.

Zweng, M. *et al* (1983) *Proceedings of the Fourth International Congress on Mathematical Education* Birkhauser, Boston.

Index